Isaac Woolworth

Our Children's Teeth, or, the Dentist to Mothers

A Manual of Dentistry for the Household

Isaac Woolworth

Our Children's Teeth, or, the Dentist to Mothers
A Manual of Dentistry for the Household

ISBN/EAN: 9783337211691

Printed in Europe, USA, Canada, Australia, Japan

Cover: Foto ©berggeist007 / pixelio.de

More available books at **www.hansebooks.com**

OUR CHILDREN'S TEETH,

OR

THE DENTIST TO MOTHERS;

A MANUAL OF DENTISTRY FOR THE HOUSEHOLD.

EMBRACING MANY PRACTICAL AND USEFUL SUGGESTIONS RELATING
TO OUR TEETH, ADDRESSED TO EVERY PERSON,
THROUGH MOTHERS.

BY

DR. ISAAC WOOLWORTH,

DENTIST.

NEW HAVEN:
PUNDERSON, CRISAND & CO
1871.

CONTENTS.

AUTHOR'S PREFACE.

In bringing this little book before the public, the Author has endeavored in a candid and benevolent spirit to offer to the reader briefly his best, and maturest convictions and judgments upon the subjects considered in his work, during a practice of over thirty years as a dentist; the conviction has been preva lent in the Author's mind, increasing with his experience from year to year, that a little work *like* this in *kind*, might be of use. And this conviction has acquired the strength almost of a posi tive *command to write*, to write earnestly, and plainly, avoiding technical forms and all ambition other than to be useful. The plan of the work being suggestive; designed for the use of every persons who retains, or desires help to retain their natu ral teeth in a safe and comfortable condition, or who may be compelled to resort to artificial substitutes. The Author, feeling conscious that the most direct way to communicate useful instruction, to the individual and to mankind, is through the enlightened co-operation of mothers, addresses the following pages most kindly and cordially to them. In doing this the Author has not aimed to give exclusive prominence to his own opinions to his own patrons, but to secure the endorsement of all DENTISTS who will unite with him in a true professional spirit, in disseminating special light among all, for the special benefit of all.

AUTHOR.

CHAPTER I.

The premature decay of the teeth and their early loss is a real and positive affliction to a large proportion of the people of our country. The miseries arising from this source are shared by so many that a good set of natural teeth has become the exception rather than the rule. The teeth are either decaying and painful, or filled—worse still, supplied with artificial substitutes. All this is an affliction, and wholly in disaccord with the economy of life and the laws of our physical existence, as well as our physical comfort.

Is it not wonderful that causes so prolific of pain, discomfort and even sickness, of sleeplessness by night and anguish by day, the dread of all ages, sexes and conditions, should so little attract the attention of writers, who, upon other subjects of interest or profit, labor so earnestly and so well to promote the happiness of our race?

It requires no argument to show that a large proportion of the physical suffering that falls to the lot of individuals and families, in one way or

another, is traceable to some condition of the
teeth. And yet so little is written for the
popular eye which merits the name or has the
semblance of authority, that it can scarcely be
said that any well defined opinions exist among
the *people,* as intelligent guides in this matter
which concerns almost every individual in civilized
life. From the infant in the cradle to the man or
woman of gray hairs, the cry is "my teeth! my
teeth!" The child, the youth, the maiden, the
mother, and the robust man alike exclaim almost
in bewilderment and despair "Oh! my teeth!
What shall I do with my teeth? I am losing all
my teeth! They are the torment of my life. I
wonder why I was made to have teeth. I wish
they were all out!" These and the like manifesta-
tions of sorrow and anxiety are patent to every
one.

It is said that some unwelcome "ghost" is in
every house. If there be any, that, like Banquo's,
will not down at any bidding, or at the most ener-
getic resort to the hundred-and-one traditional domestic remedies always at hand and recommended
—it is Toothache. What mother's heart has not
sighed or wailed at her own or her children's sorrow
from one phase or another of a raging fang? What

household has not been terror-stricken at even-tide or midnight in contemplating the dread certainty that the early dawn must drive them to that horrid chair for relief, where one intolerable pang must be substituted for another, which most *Americans* can readily appreciate.

It will be seen that special emphasis is given to the name by which my countrymen choose to be called; first, because it is the plan of this work to keep in view the teeth of Americans, both home-born and adopted, more distinctly than of other populations; secondly, because we find, at the very threshold of our labors, our own population suffering more in this regard than any other people in the world. All travelers and writers who describe in detail the habits, size and physique of different tribes of men, attest this fact. The opinions of these observers will be indicated in the course of this work, together with the opinions of eminent American Dentists, whose employment compels them to observe with much accuracy what is passing around them in their favorite and special profession. And the question has often occurred to the writer,—can anything be done to *enlighten* the *people* upon this subject. Quite as often too has it been dismissed from considerations of conscious inability to dress it up and present it so as to profit the readers for whom this little volume is written. The urgent desire that some expression of sympathy from the Dentist, as well as candid, unselfish advice, might find its way into every family in this broad land so especially afflicted with bad teeth, as well as especially blessed with good Dentists, has induced the writer to hazard the experiment of breaking the silence so long existing between Dentists and their patrons. In doing this I am conscious of the hearty support of every intelligent Dentist in our country, all of whom are looking and wishing for something of the kind, thus acknowledging the necessity.

CHAPTER II.

MISTAKEN IDEAS.

There are two very common mistakes in the minds of many persons, which I wish here to mention. One is, relying upon the Dentist to save your teeth. He can do no such thing. Whatever he may wish, or you may expect of him, he can only assist you. Inasmuch as your teeth are more important to yourself than to him, your chief reliance must, both from interest and necessity, be upon your own personal knowledge and care of them. The physician is not expected to prevent diseases from attacking you, even if you secure his services and follow his advice. He would doubtless advise you to protect yourself against those attacks. The Dentist can also give you special direction in a given case, and render you important service in the preservation of your teeth when competent service is early sought. But it frequently happens that the most valuable and faithful skill of the Dentist is nearly lost to you by your own want of vigilance or knowledge in the care of your teeth. In the following pages I hope to be able to give some general instruction, and possibly make some suggestions especially adapted to many individual cases.

The other mistake, which is far too common, is the notion that by getting rid of your teeth you escape all annoyance with respect to them. It is true that artificial substitutes of great beauty and quite useful can be supplied.

It is a great relief, sometimes, to spare a useless and troublesome limb, since it can be removed "without pain," and an artificial one of great value be substituted. But it is a great mistake to part with a set of natural teeth which can be made useful, and which also may be removed "without pain," when artificials the very best cannot supply your loss so fully as an artificial leg. This I write in high appreciation of a fine set of artificial teeth. At the same time I fully believe that the health and convenience of the body can be better secured by a substituted leg than by substituted teeth. The science and art of Dentistry are so well developed in our day, that it is hardly necessary to lose a tooth except by neglect or the misapplication of the proper means for preserving it—means embracing a great variety of palliative and protective, as well as restorative treatment, which I hope to make apparent to any attentive observer as I proceed in elucidating the theme of this work.

CHAPTER III.

It is acknowledged in every age and country, even among the rudest and most barbarous, that the teeth are of great importance in giving beauty and symmetry to the face. The most important use of the teeth is in the preparation of food for quick and easy digestion. They are also essential and quite indispensable in the perfect enunciation of language. Lord Chesterfield says that fine and clean teeth are "among the first recommendations to be met with in the common intercourse of society." Lavater remarks that "the countenance is the theatre on which the soul exhibits itself," and adds, "as are the teeth of man so are his tastes." In the French Dictionary of Medical Science is found this beautiful passage: "The teeth are the finest ornaments of the human countenance; their regularity and whiteness constitute their chief attraction. Even when the mouth exceeds its ordinary size, fine teeth serve to disguise this defect; and the illusion that results

from the perfection of their arrangement is often
such that we imagine it would not have appeared
so well had it been smaller. Observe that lady
smile, whose mouth discloses this perfection in arrangement of the teeth. You never think of noticing the extent of its diameter. All your attention is fixed upon the beauty of the teeth and the
gracious smiles which so generously expose them."

These ornaments are equally attractive in both
sexes. They distinguish the elegant gentlemen
from the sloven, and by softening the features dif-
fuse amiability over the whole countenance. " Fine
teeth," says this writer, " are more especially necessary to woman, for it is her destiny first to gratify
our eyes before she touches our souls or captivates
our hearts." The influence which the teeth exercise in the production of beauty, justifies the prominence I have assigned to them over every other attraction of the face. Let a woman have fine eyes,
a pretty mouth, a handsome nose, a well turned
forehead, elegant hair and a charming complexion,
but only let her teeth be bad, blackened by caries,
or covered with tartar or vicid accretions—let her,
in a word, exhale a contaminated breath—and the
moment she opens her mouth she will cease to be
thought beautiful. If, on the contrary, she has

small eyes, a large nose, and is even positively ugly, yet if her teeth are regular, white, and above all entire, she will appear agreeable the moment a smile comes to her aid, and she will hear the whisper so consoling to her vanity, " What beautiful teeth she has !"

Whenever nature, sometimes sparing of her gifts, fails to bestow them on the teeth, and has made them defective in form or color, great care and cleanliness are necessary to hide these defects; so that if they do not attract our regard, they will at least not fill us with disgust. Regular, white and well-formed teeth were by the ancients considered as characteristics of beauty. Jacob, in blessing Judah, says: "His teeth shall be white with milk." Solomon, in describing the church of God, compares it to a beautiful woman, and after setting forth her .graces in language that immediately brings to the mind's eye her whom Milton has described as being "the fairest of her daughters," uses the following simile: " Her teeth are like a flock of sheep that are even shorn, which come up from the washing, whereof every one bare twins." The Hebrews considered the loss of these organs as a grievous and disgraceful circumstance. Thus David uses this emphatic language: "O God, thou hast smit-

ten all mine enemies in the cheek bone; thou hast
broken out the teeth of the ungodly." Again, he
prays against the wicked judges: " Break out their
teeth, O God, in their mouth." Other references
to the teeth, as of great value, are found among the
writings of antiquity. The natives of Hindoostan,
the Brahmins, as I have been told by missionaries,
are extremely delicate in every point relating to
their teeth, often spending hours in washing and
rubbing them while addressing their deities. " This
practice is prescribed in their most ancient books
of law and divinity, and must be coeval with their
religion and government, exhibiting a curious proof
of the regard which this polished and scientific, yet
heathen, people have for the purity and beauty of
the mouth, when so simple and useful a practise is
inculcated as a law and rendered indispensable as a
religious duty. Would it not be well for us whose
knowledge and virtue are sanctified and exalted by
Christianity, if we were at least cleanly and consid-
erate in respect to gifts so ornamental to every
human face and useful to every human body?
These Brahmins, I am told, have the finest teeth
of any people in the world.

Is it not sad and humiliating that the fairest race
on the globe, the most thorough-going and thrifty

people in the world, the most painstaking and en-
ergetic, the most intelligent and economical, the
most decorative in their civilization, the world over,
should be so remiss and neglectful of the most central ornament of the human countenance, the
mouth? O mothers of America, on whom so much
depends for the adornment and comfort of your
children for life, as is involved in this matter of the
teeth; let me invite you into our company, into
communication and co-operation with dentists, who
alone can aid you in a work which must by its very
nature lie near your hearts. You can hardly be
aware how much we need and desire your co-operation in this great endeavor, the earnest effort of
every right-minded dentist, to save and restore the
teeth of your children and of the vast millions who
are to live in this land and who are destined to suffer as you and your families suffer from causes
which our united efforts can remove, or at least
greatly mitigate.

I am not charging you, in any of these things,
with conscious neglect. None are more careful and
untiring in their ministrations to the well being of
their dear ones, in everything which contributes to
their health, and to the garniture of their bodies,
than are American mothers. But if I shall be so

fortunate in these pages as to shed a ray of light, however feeble, along your pathway, your responsibilities and labors will be none the less. To aid you in these special cases shall be my endeavor, as it is my purpose. The dentist is often suspected of caring less for the safety and permanence of the teeth of his patrons than for the remuneration he receives; and he is sometimes accused of making charges greatly in disproportion to the actual expenditure of time and material. But these are not the elements of consideration in matters so important as this, and should not enter into an account in which skill and fidelity are involved.

This remark I make, not so much to defend my profession against such suspicions, as to assure you that cheap dentistry is about the poorest investment you can make; and I believe it to be almost a rule, that poor operators charge the least, not because they love money less nor the community more, but because they can acquire that little by the least expenditure of effort and the least outlay of time. Dentists of this character cannot be said to possess much ambition for excellence in their calling, or much principle as to the results of their doings.

How important it is that men whose occupation requires them to handle treasures belonging to other people, should be required to do it safely—a few reflections will show.

First, it is a duty. No man has a right to *presume* to do for others that which they cannot do for themselves, in an unskilled manner. And no man, not even a dentist, can operate successfully and skillfully on his own teeth. And whoever attempts to do this work for others without skill is doing violence to a *first principle* and *law*—to "do unto others as ye would that others do unto you." To remain disqualified, therefore, is to shirk duty and do wrong.

Secondly, a well qualified dentist is one of the most useful men in the world. Utility is one of the standards by which men estimate values; and they are frequently found sufficiently just and appreciative to reward liberally the very best specimens of our skill and our most intelligent advice. Intelligence, therefore, whether of head or hand, is our chief dependance for compensation, for renown and for justification in our own eyes and in those of mankind, while engaged in the practice of a profession such as dentistry now is. Such considerations should put every dentist upon his best endeavors to meet the exigencies of the hour, to avail himself of every facility within his reach, to keep pace with the rapid march and accurate development of our young but vigorous profession.

Thirdly, the community demands and will compel us to do just this; and the sooner we accept the situation and come into the *light,* that we *may see* and know and do accurately and advise intelligently and truly, the better will it be for our patrons and for us.

CHAPTER IV.

The infant teeth are twenty in number, or ten in each jaw, and generally make their appearance in pairs. The time of their appearance is usually from the eighth month of infancy to the thirty-sixth month, at which time they fill the jaws nearly as full as do the second or permanent set when fully developed, the shape of the jaws at this period being that of a half circle.

The great importance of these teeth renders a full account of them quite necessary. The common opinion is that because these teeth are to serve only for the few brief years of infancy, no special care of them is necessary. This is a great mistake. In the first place they are a complete set of teeth, and as positively essential in the preparation of food for digestion during the first eight or ten years of life, as the permanent set are for the remainder. And, secondly, the loss of any of these previous to the seventh year is apt to mar the beauty and regularity of the subsequent set. I have noticed that those mouths which retain their infant teeth even till the eleventh or twelfth year have generally the fairest and best permanent ones. If the removal of these teeth is by any means rendered necessary before it is nature's time to remove them, the permanent ones hasten forward while the arch of the mouth has not attained sufficient size to afford them room; and irregularities will be the result. And if these teeth are suffered to decay and blacken and die, their roots will ulcerate and often protrude through the gums; and these sharp, jagged extremities of the roots will often produce troublesome ulcers on the under side of the lips. These infant teeth should not be suffered to reach those stages of decay which make their loss necessary. They should be filled whenever they show such evidences of decay before the sixth or seventh year. Thus their vitality may be preserved till they have completed their term of service and have loosened according to nature, which is by absorption of their roots. These roots are never absorbed after the tooth has lost its vitality; they will remain till pushed out by the new tooth, and their resistance is often sufficient to divert the new tooth from its proper place, producing those troublesome irregularities so often seen. It is this faculty of absorption which so quietly and safely disposes of these teeth, and which invariably ceases when the vitality of the tooth ceases, and which attends only the removal of these infant teeth. .

There are several reasons why these teeth should not be extracted before the sixth year.

One is, if they are forced out too early, violence may be done to the rudiments of the future teeth, which are at this time in a soft formative state and easily injured.

Another is, the bones of their sockets at this tender age may be dragged out with the teeth, causing a loss of bone substance and periosteum which are never restored, except in a few instances when the periosteal membrane is preserved. Again, the order in which the temporary teeth disappear is the same as that in which they appear, in pairs, commencing with the lower centrals. It is seen at a glance that if they are extracted in any other order, the new teeth will speedily occupy the vacant places, leaving insufficient room for the remainder. Especially will this happen when the side teeth are removed before those in front, requiring tedious and often protracted operations to restore the new teeth to their positions. It is not necessary to be at great expense in filling these teeth, but that their vitality should be preserved, will not admit of a doubt.

CHAPTER V.

The diet of children for a few years after being weaned should consist largely of milk. Pure new milk contains much bone-forming and muscle-producing material. Fine bolted flour contains very little, and should be used sparingly in the diet of children while their teeth are growing. If a child acquires the habit of making its breakfast or supper of a cup of coffee or tea with a little wheaten bread or pastries of fine flour soaked in it, the effect of this depraved diet will soon be seen on the teeth. They will exhibit a pale and impoverished appearance at the points, and a blackened condition at the margin of the gums. If they are examined under this blackened surface, they will be found etched or corroded; and especially true will this be of the permanent teeth. This is probably the most prolific source of caries; since this sordid or greenish black substance surrounds the tooth at the margin of the gum, and is therefore double in amount at the point of contact with the contiguous tooth, at which point cavities usually make their appearance—between the teeth.

This condition of things is seldom found in children whose diet consists chiefly of coarse bread and new milk. Children whose appetites are not pampered by delicacies will eat with a keen relish even clear bread; and their teeth will shine like pearls. exhibiting in the first set **very little** decay.

CHAPTER VI.

THE SECOND OR PERMANENT TEETH.

These teeth, when fully developed in the adult human mouth, are thirty-two in number, sixteen in each jaw, and are divided according to the order of their eruptions, their forms and uses, into seven classes.

The central incisors, or cutting teeth, are four in number, two in each jaw, and commonly make their appearance from the sixth to the eighth year. The lateral incisors are also four in number and make their appearance after and beside the centrals.

The cuspidati, or eye-teeth, are also four in number, but do not and should not show themselves until after the jaw is supplied with three others on each side, back of the places of these teeth, which in nature's order are occupied at this time by the same class of the first set. The bicuspids, or two pointed teeth, as their name implies, are eight in number, two on each side of each jaw. These teeth occupy the places formerly occupied by the molar or double teeth of the first set. Next are the first or six-year-old molars, or grinding teeth, called by dentists, six-year-old molars, because of their eruption at that age. These teeth require special notice, because they appear usually previous to any others of the permanent teeth, and are therefore found in company with the infant set, with which they are often confounded; so that many people, supposing them to belong to this set, suffer them to decay away under the mistaken impression that they will be followed by new ones. This is never the case. Once removed, they are removed forever. These teeth are of so much importance in mastication as to require some special notice, which will be given further on in the chapter.

The sixth class are the fourteen-year-old molars, and are the second double teeth which present themselves behind the class just described, from the twelfth to the fourteenth year of age, and are heavy strung teeth, capable of enduring a vast amount of service if properly preserved.

Lastly may be considered the "dentes sapientia," or " wisdom teeth," four in number, which make their appearance, not without a vivid impression upon the memories of most persons, from the eighteenth to the twenty-first year. I do not know as they bring along with them either wisdom or patience, except it be "through suffering"—for most persons have good cause to remember the " cutting" of their wisdom teeth. Great suffering often arises from this source which requires these teeth sometimes to be removed to obtain relief; and this not unfrequently is far from being a desirable alternative, especially when, as often happens, the parts adjacent have become swollen.

These teeth—these thirty-two flinty organs of the mouth—are more sordid and compact in their structure than any of the numerous tissues or formations of the body, and are destined to outlast them all. They are frequently found in a fair state of preservation after all other portions are literally "turned to dust."

That these organs are *made to endure* the repeated shocks and constant wear and friction of reducing the hardest substance of man's proper diet during three score years and ten, is shown not only by their composition and structure, but by the *entire absence of any provision for their reproduction.* Only once in a lifetime are such gifts as these vouchsafed to man. This fact, so significant of value and durability, may well suggest to us the importance, in the animal economy, which the giver attaches to them; and how imperative and binding is the spirit and intent of a law whose infraction is visited so promptly and permanently by its penalties. Such prodigality, and such disregard of consequences so severe and lasting, is seldom witnessed or so strikingly illustrated as in the innumerable examples of the teeth of Americans.

Submission with an amiable grace to the natural encroachments of time, the decreptitude and spoliation of age, is both heroic and wise. But to suffer the stealthy footsteps of any insiduous destroyer, havoc in hand, to approach these inestimable gems, is neither wise nor necessary. Did those agencies which attack the teeth like a devouring flame, herald their approach by the more positive testimony of the senses, the instincts of animal life might suffice to teach us prudence in the presence of danger. But when the causes which conspire to rob us of a lifelong enjoyment of nature's choicest gift, are not only strewn along the whole period of life, but are latent at the very springs of existence, it becomes a subject of study and philosophy to every parent and person of reflection in the land; how can these effects be averted, or how can we disarm them of their power of evil? To disarm these insiduous ene-

mies of our peace, or to intrench ourselves behind
the substantial bulwarks of prudence and care;
to use our reason and philosophy; to call to our
aid intelligent and skillful dentistry, is all that
any grown man or woman can do to arrest those
processes which denude, disintegrate and destroy
the teeth.

But inaccuracies and degeneracies, mental and
physical, as well as improvements and perfections,
are also transmitted. We inherit alike the proportions and powers of both mind and body as well as
our temperaments and powers of endurance and
our qualities of soul. The teeth, as any observer
may know, share in the distribution of this primal
law of our being. Analogies of this kind may be
found in every department of culture. The florist
may change and improve the size and tints of his
flowers, as well as multiply their petals by cultiva-
tion and nutriment. The gardener may improve
the quality and increase the quantity of his fruits
and vegetables. The herdsman may also heighten
the beauty and power of his beast by attention to
nutrition and the laws of propagation; while the
reverse of all this may happen through disregard of
those laws. It is a maxim, old as the history of
our race, and true as the sacred precepts of the in-

spired book, that "the fathers have eaten sour grapes and the children's teeth are set on edge." If mothers feed upon "slops" and delicacies to the exclusion of more nutritious aliment during all the months of pre-maternity, not only will their own teeth become diseased, if not destroyed; but the teeth and bone and muscle-producing quality *in their children* will certainly become impaired, whence it will be easy to predict for them all the suffering arising from this inherent degeneracy.

CHAPTER VII.

The same result ensues from using too much fluid in the form of drinks of any kind, especially if taken while eating. Too much watchfulness can scarcely be exercised at this time ; and children, unless limited in this particular, will almost surely acquire a habit which will result in great injury to their teeth. It will not do to reply that "I drink only water;" fluids of any kind dilute the gastric juice, a special secretion found in the stomach and indispensable to digestion. Too much fluid in the stomach retards also the process of digestion as all surplus must be taken up by the absorbents before the process can go on. Milk taken into the stomach for food does not digest at all until it is changed into a solid in the form of curd of more or less firmness.

In the description, given in a former chapter, of the permanent teeth, I promised a more extended notice of one class, namely: those called the six-years-old molars. Special attention is called to these teeth on account of their importance and position in the mouth. These teeth, from their central position on each side of both jaws, are made to receive and sustain the greatest force in the labor and luxury of mastication, and are therefore of first importance in this process. The loss of them not only greatly impairs the apparatus for breaking down and reducing food to a proper state for the stomach; but their preservation is indispensable to that complete antagonism or articulation so essential to an entire set of teeth. If they are suffered to decay or are removed, their absence from this central position causes the contiguous teeth to stray from their natural position, and to incline forward or backward, so that their crowns or grinding surfaces do not fairly meet the opposite teeth.

Furthermore, these teeth erupting as they do before the infant set disappear, are much exposed to decay from the decaying condition of those infant teeth. They should at this time be thoroughly filled with gold, and *early*, because the cavities reach the pulp or nerve much more quickly in children than in the adult. I may here add that the walls of all teeth are much thinner, and the interior portion or nerve cavity is much larger at this age; both of which features change in this particular from year to year until old age, when these nerve cavities wholly disappear, and the tooth becomes solid. This may appear surprising, but I have met with a few such cases; and only a few, because but few teeth in our day can boast of so venerable a history.

This fact furnishes a strong argument in favor of filling teeth, when the distance between the bottom of a cavity properly filled and the living nerve is constantly increasing, removing the vital part farther and farther from danger.

It is also worthy of remark that in these days of degenerate teeth the proper time to begin to look after them is very early, for it frequently happens that the infant set require the aid of the dentist other than in their removal.

But the second or permanent teeth are those about which persons wish to know most. To develop a sound and regular set of teeth, and to preserve them, is an object worthy of some care. But it is regarded by many as almost hopeless; and no wonder, when we consider the prevailing liability of teeth of Anglo-Americans to decay, as also the amount of general ignorance and indifference in this regard, and the wonderful dearth of published instruction on the subject for the people's reading. I may safely venture the assertion, that there is no single interest of one-tenth the importance to every man, woman and child in this nation, that has not its special advocates and teachers by scores and hundreds; while the pens of eminent dentists all over the land are in effect *dumb,* so far as diffusing the knowledge they possess among the people. Much oral instruction is doubtless imparted by the benevolent dentist to the patient in his chair; and this might be sufficient if all talk were gospel, which it were hardly prudent to assert. The cause of this prevailing silence, and therefore inevitable darkness, may be found in the fact that such men find full employment in repairs, and have no time to instruct, or are too modest to proffer advice unasked, or value too highly such choice possessions, to give them away to persons who might treat them as broken toys.

How many myriads of teeth annually lost might be saved by a full and intelligent dissemination of knowledge already possessed by the intelligent dentist! It is estimated by Dr. John Allen, of New York, that 20,000,000 of teeth are annually lost to the American people, which this knowledge and the improvements in surgical and operative dentistry might save, to do honest service for years. This is not necessarily the sad inheritance to which our children are born. With a view to increasing this knowledge and adding interest to a subject of vital importance to so many millions of my country-men, and also to honor my distinguished friend, who is at the same time a real friend to the people, I give my readers some of his thoughts and contribu-tions on the subject.

The following chapter, the product of a fertile pen and generous heart, was read before the Amer-ican Dental Association, in Cincinnati, August 2, 1867, and entitled " The Physical History of Vari-ous Nations of the Earth, with Special Reference to their Teeth." *

To live without teeth, save such as are artificial, is wholly unnatural, and, as a rule, unnecessary.

Equally unnatural and unnecessary is it that any one should receive an originally imperfect set, when Nature in the distribution of her gifts is no re-specter of persons, but has amply provided that every organized being may enjoy equality with every other, upon equal conditions.

If one child be trained in the nurture and admo-nition of the Lord, and enjoys therefore the bless-ings of obedience, how can another expect equality with him, or claim exemption from the results of a

*See Chapter VIII.

broken law, who lives in habitual violation of its precepts? If, for example, a given amount of lime or lime phosphates be taken into the system, to insure the development of fair proportions in the bones, muscles and teeth, how can these proportions and qualities be expected when these substances are almost wholly wanting? Says Dr. John Allen : " It has been estimated, and is a fact, that a person of ordinary stature who uses for bread common wheaten bread made of bolted or fine flour, is deprived, every year he uses it exclusively, of twenty pounds of the very material of which bones and muscles and teeth are composed."

If this be true, of which I have no doubt, it is easy to account for the constantly diminishing stature of American men and women from generation to generation, as well as for the degeneracy in the size and structure and durability of their teeth.

That this is true, is inferred in part, at least, from the oft-repeated assertion of multitudes of persons whose teeth exhibit this degeneracy, that " my father died at seventy or eighty years and all his teeth were sound," or that " he had double teeth all around." These traditions, though not strictly true, indicate a better physical condition in our plainly fed grandsires than in their pampered luxury-nurtured children. That anybody ever had double teeth in front is a gigantic fib; but the statement proves that this wonderful progenitor had teeth which resisted decay. The appearance of being "double" was the result of wear from long and valuable service, and from peculiar occlusion.

CHAPTER VIII.

THE PHYSICAL HISTORY OF VARIOUS NATIONS OF THE EARTH, WITH SPECIAL REFERENCE TO THEIR TEETH.

By DR. J. ALLEN.

Read before the American Dental Association, in Cincinnati, August 2, 1867.

Having spent some thirty-eight years in Dental practice, I have often been asked these two questions: first, "are not the teeth of the people of this country worse than those of other nations of the world?" And, second, "what is the cause of so many bad teeth in America?" These are two important questions involving the welfare of some thirty millions of inhabitants. In order to answer them satisfactorily, we have found it necessary to examine the physical history of mankind, in order to compare nations with nations in reference to their teeth, taking into consideration their food, habits, customs, climate, etc.

In prosecuting these researches we find there are many nations whose teeth remain sound, even to old age, and it is as rare for them to lose a tooth as it is an eye or a limb. While in this country it is estimated that there are more than twenty millions of teeth lost annually from decay. And yet we find that the same general physical law which provides for the building up and sustaining the human structure, prevails among all nations, and that the divine architect of man has furnished an abundant supply of materials for all parts of the system. The body of man, with all its different parts and organs, is composed of only a few simple materials. These are combined in certain proportions, in order to give strength and utility to the whole structure. These materials are component parts of his food; and although the nutrient substances used by the inhabitants of different parts of the world appear quite dissimilar, yet the food provided for them in various countries possesses the same general nutriment properties and chemical constituents everywhere that are essential for the human organism.

We will now proceed to notice some of the historical evidences which go to establish the fact that Americans, as a whole, have worse teeth than the inhabitants of other nations. In portions of Europe, where the people, like the Americans, discard a large portion of the mineral element from their food, they also have bad teeth; but among the Peasantry, and also in those sections where the inhabitants do not change the proportions of the mineral constituents of their food, they have good teeth.

But let us turn to the historical accounts of other countries where bolting cloths are not used for this purpose.

In Prichard's Researches into the Physical History of Mankind, he says: "The Albanians of Lesser Asia live principally on milk, cheese, eggs, olives and vegetables. Sometimes they bake bread, but often eat their corn or maize boiled." Hippocrates says they are very strong and muscular, have oval faces, a ruddy color in their cheeks, a brisk, animated eye, a well proportioned mouth, and fine teeth. In central Africa, north of the equator, Prichard says "the Mandingo tribes have the barbarous custom so common among the Pagans of Africa, of filing their teeth to a point."

In eastern Africa, among the different races of Abyssinians, we have the following description by this eminent author: "Their countenance is full without being puffed, their eyes are beautiful, their mouth of moderate size, their lips thick, their teeth white, regular and scarcely projecting." Among the races of people inhabiting Nubia and other countries between Abyssinia and Egypt, Burck-hardt says: "They are a handsome and bold people, of a dark brown complexion, with beautiful eyes and fine teeth." In the western parts of South Africa, comprising the Congo Empire, Proyart, who has graphically described it, says "the negroes are well made, very black, with white teeth and pleasing countenances." In Dr. Oldfield's ethno-graphical researches in the interior of Africa, among the Felatahs, he says: "The color is light brown, features regularly formed, handsome mouth, thin lips, with teeth as white as ivory."

We will now pass into Asia, and there among the mountain tribes of Dekham, in India, Dr. Max-well says: "The Khonds are a dark race of men, straight, well limbed, and free from obesity, which makes them have a tall appearance. Many of the men have a pleasing expression of the counte-nance. Generally, however, the nose is flattish, the cheek bones high, the face round, the lips and mouth large, displaying fine teeth. The country produces rice, and most of the vegetables which are

common in Europe." Among the Turkish tribes
of Kiptschak, the Tartars of Kasan, says Erman,
"are of middle stature and muscular, but not fat.
Their heads are of an oval shape, their countenances of fresh complexion, and fine, regular features ;
their eyes, mostly black, are small and lively ; their
noses arched and thin, as well as their lips; their
hair is generally dark, and their teeth strong and
white." We will now pass to that part of Asia between Hindoostan and China, where we find, according to Finlayson, that the Siamese blacken
their teeth and redden their mouths with a masti-
catory of lime, catechu and betel, which gives them
a disgusting appearance. Baron Larrey, who is
well known as an eminent author on physical subjects, says : "The inhabitants of Eastern Arabia
are somewhat above the average stature, robust
and well formed. Their countenances oval, and
copper colored, the forehead broad and elevated,
the eyebrows black and bushy, the eye dark, deep-
seated and quick, the nose straight and of moderate size, the mouth well shaped, *the teeth beautiful
and white as ivory.*" "In Egypt," the same author
says, " the surface of the jaws of the Arabs are of
great extent and in a straight or perpendicular
line. The alveolar arches are of moderate size, and

supplied with very white and regular teeth, the canines especially project but little." The Arabs eat little and seldom of animal food.

We will now pass to a group of islands situated in the great Southern Ocean, between the eastern coast of Africa and the western shores of the new or American Continent. This group of islands received, from Captain Cook, the name of the Society Islands. Mr. Ellis, who spent some six years among the inhabitants of Tahiti as a missionary, and had ample opportunity of observation, says: "These people are above the middle stature; in physical power they are inferior to the New Zealanders. The mouth of the Tahitian, he says, is well formed, though the lips are sometimes large, yet never so much as to resemble those of the African. The teeth are always entire, except in extreme old age, and though rather large in some, they are remarkably white, and seldom either discolored or decayed."

Mr. Anderson, who visited New Zealand with Capt. Cook, says: "The nations do not exceed the common stature of Europeans, and in general are not so well made, especially about the limbs. Their color is of a different cast, varying from a pretty deep black to a yellowish or orange tinge, and their features are also various, some resembling Europeans. Their faces are round, with full lips, their eyes large, hair black, straight and strong. Their teeth are commonly broad, white and well set." Another writer, Captain Fitzoy, in describing the people of New Zealand, where he speaks of their teeth, says: " They are like those of the Tuegians, and, at the first glance, remind one of those of a horse. Either they are all worn down, in old persons, canine, cutting teeth and grinders, to an uniform hight, so that their interior texture is quite exposed, or they are of peculiar structure," undoubtedly the former. The natives who live near the hot sulphurous waters on the borders of the lake of the Roturna, have the enamel of their teeth, especially their front teeth, yellow, although this does not impair their soundness, and is the effect, probably, of the corroding qualities of the thermal waters. To the eastward of the Society Islands, in the South Pacific, are the Gambier Islands. They are inhabited by a people fairer than the Sandwich Islanders. The average hight of the men is about that of Englishmen, but they are not so robust. In their muscles their is a flabbiness, and in the old men a laxity of integument which allows their skin to hang in

folds on different parts of their body. They have
an Asiatic countenance, the teeth in the fourth
class especially are not remarkable for evenness or
whiteness, and seem to fall out at an early period.
With reference to these physical characteristics.
Dr. Prichard says: "Two causes may be assigned:
the nature of their food and their indolent habits."

We will now pass to Easter Island, which is
situated perhaps the most remote from the great
continents of all inhabited islands on the globe.
Captain Beechey has given the following physical
account of the inhabitants. He says: "They are
a fine race of people, especially the women. They
have oval countenances, regular features, a high and
smooth forehead, black eyes and fine teeth.

Next, let us take a view of the Samwan group of
islands, situated also in the Pacific ocean, in latitude thirteen and fourteen degrees. The inhabitants of these islands are strong, vigorous and well
proportioned. Their features are all referable to a
common type. This type is thus minutely described: "The nose is short and wide at the base,
the eyes are black, and often large and bright, the
forehead narrow and high, the mouth large and
well filled with white and strong teeth." These
islands abound in pigs, dogs, fowls, birds and fish,

and likewise in cocoa-nuts, guava, banian trees
and sugar canes. Belonging to another group, in
the same ocean, are the Tarawan Islands. The
people of this group differ from those above described. They are of middle size, their color is
dark copper, their hair is fine, black and glossy,
the nose slightly aquiline, the mouth is large, with
full lips and sound teeth.

Vanikoro, another group of these islands, is also
in this great ocean. The sea coast is inhabited by
a black race, who cultivate the taro, iguamas, ba-
nanas and the kava. "The inhabitants," says Dr.
Urville, "belong to the black race of the great
ocean approaching to that of proper negroes.
They are generally small, their countenance has a
singular resemblance to the ourang-outang, the
eyes are large, and deeply set, resembling in form
and color, those of the negro. The lips are large,
the chin small and the hair crisp. The use of
the betel root destroys their teeth, and gives them
a red tinge round the mouth. The women are
horribly ugly, the old men are bald." Next we
will proceed to the Archipelago, of the Fiji or Fejee Islands, which lie to the eastward of those
above named, and are situated between fifteen and
and nineteen degrees of south latitude. This is a

large group of islands, many of which are inhabited. The largest of this group is called the Great Viti. The people of this island are called Vitians. They are tall, well made, active and muscular. Their faces are broad, noses large and flat, large mouths, thick lips, and sound, white teeth.

The Fejeeans are generally above the middle hight, and exhibit a great variety of figure. Their complexion is between that of the black and copper colored races, although instances of both extremes are to be met with, thus indicating a descent from two different stocks. The faces of the greater number of the Fejeeans are long, with large mouth, good and well set teeth.

Leaving the Fejee Islands, we will pass to the coast of Australia. Captain Wilks, who was sent by out by the government of the United States on an exploring expedition, says: "The natives of Australia differ from any other race of men in features, complexion, habits and language. Their color and features assimilate them to the African type, their long and black silky hair has a resemblance to the Malays; in their language they approximate more nearly to our American Indians, while there is much in their physical traits, manners and customs to which no analogy can be traced in any other people. Their color usually approaches chocolate, a deep umber or reddish black, varying much in shade in different individuals. The cast of the face is between the African and Malay, the forehead usually high and narrow, the eyes small, black and deep set, the nose much depressed at the upper parts, the cheek bones are high, the mouth large and furnished with strong, well set teeth."

We will next direct direct your attention to the American races or tribes of this continent. Dr. Morton, who has published a very popular work on American skulls, says : " It is an old saying among travelers, that he who has seen one tribe of Indians, has seen all. So much do the individuals of this race resemble each other that notwithstanding their immense geographical distribution, and those differences of climate which embrace the extremes of heat and cold, there is a remarkable identity of physical characteristics throughout this whole race of people. All possess, alike, the long, lank, black hair, the brown or cinnamon colored skin, the heavy brow, the dull and sleepy eye, the full and compressed lips, and the salient but dilated nose.

Without following this author through his details of the physical characteristics of the American races, we will pass at once to his records of their teeth. He says: "The cheek bones are large and prominent, the upper jaw is often elongated, but the teeth are for the most part vertical. The lower jaw is large and ponderous; the teeth are also very large and seldom decay, and few present marks of disease, though often worn by the mastication of hard substances."

With reference to the nations of the western coast of North America, we have the following record from Capt. Cook and Mr. Anderson:

"The visage of most of them is rather round and full, and sometimes also broad, with high prominent cheeks. The nose flattened at the base, the forehead is rather low, the eyes small, black and languishing rather than sparkling, the mouth round with thickish lips, the teeth well set but not remarkably white."

We will now direct attention to the nations of Chili, of California, and to those of the country near the Baie des Francais, who are of the Kolushian race. In the historical account of these people, by Mr. Rollin, we have the following: "They have rather a low forehead, black and lively eyes,

nose of a regular shape and size, rather wide at the
extremity; lips fleshy, a mouth of middle size, fine
and well set teeth." We will also notice the Peru-
vian nations. The physical characteristics of these
nations in general are described by Dr. Orbigny.
He says: "Their features have an entirely peculiar
cast, which resembles no other American people
but the Mexicans. Their head is oblong from the
forehead to the occiput, somewhat compressed at
the sides. The forehead is slightly arched, short,
and falling a little back. Their face is generally
broad, approaching to an oval form; their nose
prominent, long and strongly acquiline; the mouth
is larger than common, though the lips are not very
thick. The teeth are always beautiful even in old
age." Dr. Orbigny says the mountaineers in South
America are generally short, while the inhabitants
of the plains are tall. "The Aroucans are a square,
stout set of men with robust limbs, but without
obesity; their joints large, their hands and feet
small. Their heads are large in proportion to their
body; the countenance full, round, with prominent
cheek bones, large mouths, but thin lips. Their
teeth are good and remain sound in old age.

"The aboriginal nations of Eastern Patagonia,"
says Capt. Fitzroy, "are a tall and extremely stout

race of men. Their color is a rich, reddish brown. The head of the Patagonian is rather broad, but not high; the mouth is large and coarsely formed with thick lips. Their teeth are usually very good, though rather large, and those in front have the peculiarity of being flattened, solid, and showing an inner substance." The following is an extract containing a description of the Pesherais, a people who inhabit one of the islands of the Magellanic Archipelago. (This extract is taken from an account of an exploring expedition sent out by the United States government.)

"These people are not more than five feet high, of a light copper color. They have short faces, narrow foreheads and high cheek bones. Their eyes are small but unusually black. Their nose is broad and flat, with wide-spread nostrils, mouth large, teeth white and regular." Dr. Orbigny, in describing another tribe of the South American Indians, (called the Botocudos) says: "They wear for ornaments, collars or strings of human teeth." This is an evidence of their soundness and beauty. In the northern division of South America we have the following physical description of a people called the Chaymas. Humboldt has given the following description of them:

"The countenances of the Chaymas, without being hard and stern, has something sedate and gloomy. The forehead is small and but slightly prominent. The eyes black, sunken, and very long, the mouth wide, with lips but little prominent, has often an expression of good nature. The nose and nostrils resemble those of the Caucasian race. "The Chaymas," says Humboldt, "have fine, white teeth, like all people who lead a very simple life."

Having taken a cursory view of the physical characteristics of several other nations of the world, with special reference to their teeth, we will now return to our own country. We have here a mixed population from various parts of the world, who have become so assimilated in habits, manners, customs, mode of living, etc., that the historian would recognize the same general physical characteristics of the people throughout the United States. But how different would be his record in reference to the teeth of the Americans at the present time from those nations herein referred to. He would tell you that very many of the people of this country have narrow, contracted jaws, with crowded and badly decayed teeth. And in his statistics he would announce to you the startling fact that twenty millions of teeth are annually lost by the people of this country. From the evidences which we have endeavored to bring before you, it will be seen that the teeth of the people of this country are far worse than of any other here described. Mark the words of Humboldt when he said: "The Chaymas have fine teeth like all people who lead a very simple life." It will be observed that in these historical researches, there is no evidence that the nations Humboldt alluded to attempted to improve their food by changing the proportions of the different constituents which the Creator has duly apportioned for the building up of organized beings. But, on the contrary, those nations use their food in the most simple forms, with all the constituents which nature placed there for the use of man. Another important fact in the history of those nations who have well developed jaws and teeth, should be also noted. It is this: they have plenty of exercise in the open air, which enables them to appropriate the different constituents in their food to the various parts and organs of the human system. From these different nations, therefore we may learn some valuable lessons on the subject of the teeth. Although they have no Dentists nor Dental literature, (for they need none) yet they learn much, as we may, from Nature, which will be found to tally exactly with true science.

Now let us turn again to our own records and see how widely we have departed from some of those physical laws which have been established by Omnipotence for our well-being. We have vainly attempted to improve our bread (the staff of life) by changing the proportions of the mineral element in the flour we use, by bolting the most of it out and discarding it. Look for a moment at the gigantic scale upon which it is done in this country. According to our national statistics of 1860, there were in the United States 13,868 milling establishments for the manufacture of flour and meal, requiring 27,626 men, at an annual cost for labor of $8,721,391. Thus, you see, the number of men, mills, bolting cloths and dollars that are employed in this great *improvement* (?) devised by man for changing the proportions of one of the most important constituents in the staple article of food in this country. The result of ignoring this mineral element from the staff of life is, undoubtedly, to a great extent one of the most prominent causes of this national calamity, that sweeps from the population more than 20,000,000 of teeth every year. The potter cannot make the bowl without the clay, neither can good teeth be formed without a due proportion of lime, which is abundantly provided for our use upon the outer portion of the grain, and in rejecting this portion of the cereals we virtually refuse to use the requisite materials of which the teeth are formed. We also deprive ourselves of a due proportion of atmospheric constituents, especially in our crowded cities. And also of the requisite amount of exercise to promote vigorous health and good constitutions. If we would be instrumental in doing more good in our profession, let us do all in our power to diffuse these important truths among the people

CHAPTER IX.

THE RELATIVE IMPORTANCE OF THE SECOND OR PERMANENT TEETH.

This set of teeth, as their name indicates, becomes the object of special consideration, inasmuch as they are the only provision of nature for the purposes for which they are given, during the whole period of life nearly, after the sixth year. At this time the infant teeth, having accomplished the work assigned to them, begin to disappear. Then man is put upon his good behavior in the judicious husbandry of his last natural gift of the kind—not knowing how long he may want them, nor how sorely he will regret their loss when driven to the trying alternative of wearing false teeth or going without.

To preserve them from decay and pain, and from ultimate dismemberment and ruin, so common in this country, has become a subject of almost universal anxiety. How shall I preserve my teeth, is the the anxious inquiry of thousands who go reluctantly as "lambs to the slaughter," to some of the many hundreds of dental offices for "repairs" or advice, but too often too late for the most efficient service. This is well, the best thing that can be done at the time; but if the tooth has become painful, it is evident the visit has been too long delayed. But whatever may be the fate of this first offending tooth, the visit will doubtless be timely with respect to the remainder; but very much trouble and pain might have been avoided by a little more knowledge of the teeth.

We will suppose this first trial, this first admonition, to be at the age of fourteen years. There will be at this period, in the mouth, twenty-eight teeth, which are to accompany you through life. They are among the choicest gifts of your being. To their soundness and cleanliness you are to owe a large portion of your physical happiness each day of your life; and from their diseases and loss you must of necessity suffer the greatest proportion of physical disquiet which falls to your lot. It is of very great importance while these teeth are coming that they be often and carefully examined, to know whether they are coming regular, and whether cavities exist or no. Very much trouble can be prevented by early attention; and the dentist can at this time render you essential service, even if it is not necessary to operate. The relative importance of this set of teeth to those that preceded them is as the whole of life to a few years.

To preserve them for many years is one of the most important considerations of the spring-time of life, because no system, so organized that teeth are required for the purposes of nutrition, can possess good health without them. Nor can teeth which are decaying or diminishing in number perform this work perfectly, when every one of them, in a perfect set, is made to meet the corresponding one of the opposite rank in exact occlusion.

Every mouth is thus furnished with a complete mill for the grinding of food. This mill being present in each mouth is evidence that something is to be ground, and that the proper diet of man requires this process, and that it should not reach the stomach without it. Any mode of cooking which absolves us from this necessity is an injury to all otherwise wholesome diet. And any condition of the teeth which deprives us of this luxury in eating, is a positive injury to health as well as to the teeth themselves, and if persisted in, will surely destroy them. It is rebellion against an essential law of physiology which requires the exercise and use of every organ or function which we wish to preserve. Let a person who possesses good teeth live upon food so prepared that it can be eaten without teeth, and he will soon pay for such folly by the loss of every tooth. Let him sit at an American table and note the effect of American cooking on his own and his children's teeth, to prove this doctrine.

These effects may be more safely demonstrated by noticing the result of unnatural feeding in animals. If a bullock be put to fattening upon sour distillery slops, a very few months only will elapse before he will have to be slaughtered to save him; because he will have no teeth, and he cannot live long deprived of solid food, either hay or straw, and the ability to eat it. Lap dogs and poodles are almost the only specimens of their race which lose their teeth, because they are often fed on sweetmeats and other luxuries of the table. A cow accustomed to feed on the honeyed clover, and to yield the "pure milk" of poesy and song, if put to the rascally production of "swill milk," can never again "lead an innocent rural life, because this constrained and cruel mode of living destroys her teeth and *otherwise dismembers her body.*"

CHAPTER X.

It is quite evident that teeth and other parts of the body cannot be developed or sustained upon food which does not contain the elements of which the parts of the body are composed. The fine flour of which most of our bread is made contains only 30 parts in 500 of these materials, while whole grain, such as wheat, contains 85 parts, and the bran separated from it by bolting contains 125 parts (in 500) even after the skin is removed.

It is a want of these qualities in our daily food, and the perverse methods of preparing it, which sap the *foundations* of development and nourishment among our people, and to which this whole-sale disintegration and destruction of the teeth may be attributed. Professor Johnson, in the Patent Office Report for 1847, has made the following curious but practical observations. In examining wheat and flour as to the amount of nutrient or muscular matter, the fat-forming principle, and the bone and saline material contained in grain in different states of preparation, he found that in 1,000 lbs. of whole grain there were, of muscle-producing material.

	Fat principle.	Bone and saline.
156 lbs.	66 lbs.	170 lbs.
In fine flour, 140 "	20 "	60 "
In bran, 694 "	60 "	70 "

In the same work we find the difference in the weight of a barrel of flour, without the bran, and when only the outer coating of the wheat is taken off. He says, "the weight of the bran or outer coating in a barrel of flour, constituting the offal, is only 5½ lbs., while the ordinary weight of the offal is from 59¾ to 65 lbs. in every barrel of flour, which amounts to a loss of 40 lbs. of earthy constituents which might be reserved in every barrel of flour."

Our object in this exhibit is to awaken thought and encourage a more enlightened and systematic regard to diet with reference to teeth. If we succeed in this, our object will be gained, and we shall be satisfied in our choice of a difficult and trying, but necessary, profession, and shall feel that we have not lived altogether in vain.

CHAPTER XI.

CLEANLINESS WITH RESPECT TO THE TEETH.

Cleanliness is a virtue at all times, and fortunately is enjoined upon all not only by the requirements of civilization, but of health. But with respect to the teeth, society is not so exacting. It is true that clean and white teeth are a very positive necessity with many, and any neglect in this duty of the toilet is to them a gross violation of their standard of propriety. Yet bad and even uncleanly teeth do not exclude one from good society, though certainly they are no passport to refined social fellowship. And I think it would contribute much toward a better condition of the teeth in American society, if the laws of social etiquette were more stringent in this particular. No person of ordinary taste would for one moment think of appearing in society in tattered garments, with disheveled hair, unwashed and untidy. Many must even be perfumed and powdered and adorned in varied splendor, from tinsel to diamonds, to make themselves presentable, whose *teeth* by no means exhibit corresponding care!

I do not know how many besides those of my own profession will endorse this sentiment; yet I believe there are many more admirers of fine clean teeth than there are dentists.

I should value very little the taste or faithfulness of a dentist, who did not *impress* upon his patients the value and necessity of cleanliness with respect to the teeth. It might not be policy, as the world sometimes defines it, for the shoemaker to caution people to take special care of their shoes, or for the doctor to go frantic over people's recklessness in regard to health. But it is good and wholesome counsel and will injure nobody, for the dentist to insist upon cleanliness of the teeth, and to instruct every one of his patients how to accomplish it.

This chapter would be incomplete if it did not give quite full directions on this point.

Wash the teeth thoroughly with water after each meal, using a quill tooth-pick between each tooth and its neighbor, to clear away any particles of food remaining between them, which would soon be rendered acrid, and if the food contain sugar or starch, the temperature of the mouth would soon convert these into acid. This acid attacks the limy portions of the tooth and dissolves them, and thus they are eaten in holes at points of contact and in their indentations on the crowns.

Another mode recommended by some eminent dentists is, to pass between the teeth a thread of floss, silk, or rubber. The rubber has one merit above anything else, only, that it can be passed between any of the teeth, however closely they may approximate.

A little fine soap on the brush might be used to great advantage, as the saliva in these conditions has most invariably an acid reaction. In short, all dentifrices, to control this acid condition, should contain some of the milder alkalies, such as magnesia, or super-carbonate of soda, or soap. Prepared chalk ("creta preparata" of the shops,) is still better. And when streaks or stains appear at the margin of the gums, a little fine grit should be used occasionally upon the end of a flat pine stick with a blunt square end, to remove these stains and polish the enamel. The resort to the dentist at short intervals to have these stains " scraped off," is not judicious. The dentist should be required to *scour* off these stains, not

scrape them off. Thus he will accomplish two
things for your teeth—cleanliness and smoothness.

It might seem paradoxical that a dentist should
recommend such attention to the teeth as would
postpone the necessity of his services in repairing.
But the dentist much prefers to find the teeth
clean and the gums hard and healthy when he
operates; and then he is sure his operations will
do him double credit, and the patient will realize
manifold more in having the teeth kept nice and
clean after filling; and it is better too in a pecuniary sense for the patient, and quite as pleasing
to the dentist.

Powdered charcoal is often employed as a dentifrice; but many eminent dentists have spoken
against its use, because they think its mechanical
effect too severe on the enamel. And then, too,
this substance, when reduced to a very fine powder,
insinuates itself into the pores or open mouths of
the mucous follicles, along the margin of the
gums, producing a permanent blue line around
and between the teeth. This I have seen, and the
stain is as indelible as gunpowder shot into the
face, or as tattooing with India-ink. In my judgment, charcoal, properly prepared and judiciously
used, is one of the best of dentifrices. It should

not be reduced so fine as to permeate the gums, and should be rubbed upon the teeth with a stick lengthwise of the tooth, one at a time, and not in that wholesale manner in which tooth polishes are often used, as one would polish a stove. Don't be in a hurry; do one tooth at a time; and my word for it, you will be paid for your time.

The gums never need scrubbing when the teeth are kept clean ; and they will often bleed on apply- . ing the brush. This should not be; the gums should never bleed, except as a relief for engorgement. That turgidity often seen along the margin of the gums is commonly caused by some extraneous deposit or some unnatural roughness on the teeth underneath their free edges. It is only necessary to remove these points of irritation to restore them to health; some astringent wash will be useful on such gums. The habitual loss of blood from the gums is sure to result in a loss of substance of the gums, and even of the sockets of the teeth themselves. This is true of all those cases of loosening and falling out of the teeth while yet sound. The gums in a healthy state are not red, as if engorged with blood, but are a light pink, and in young persons almost white, and bleed a little less readily than the skin.

When the gums bleed on the slightest touch, there will be found under their edges portions of a hard crystalline substance called "tartar." This keeps them in a constant state of irritation, and should be removed with an instrument, as it cannot be removed without it. No amount of brushing will remove tartar when it once becomes hard.

This tartar collects in greater abundance in the lower than in the upper teeth, especially those in front. This kind of deposit does not cause the teeth to decay; but it destroys them quite as effectually, and should not be allowed to remain. A great proportion of the teeth which are lost are lost from this cause solely; as most of those teeth which fall out from this cause are entirely free from decay, and their loss may be attributed solely to want of care in these particulars.

As soon as a particle of this substance becomes attached to a tooth it forms a nucleus to which other particles aggregate, till all undefended points are covered, inside and outside, with a hideous crust almost as hard as a stone. Unless this is removed and kept off, it will force away the gums even down to the points of the roots. And its effects by no means cease with the changed conditions of the gums, but result often in the entire destruction of the alveolar ridge in which the teeth are planted.

Many interesting experimental operations have been made by a few exact and thoughtful operators and skillful men in our ranks to restore the gums which have been forced to retire, leaving the necks of the teeth and all the portions designed to be covered, bare, and the teeth swaying to and fro at every movement of the tongue and lip. and to build these denuded gums to their normal conditions. And I am not without hope that this troublesome condition of so many mouths will yet be controlled and treated successfully. I would therefore call the attention of my readers to this great destroyer of human teeth and contaminator of the human breath, for the sole object of saving from destruction the natural organs; because so many persons are misled by the heretofore opinions of physicians and dentists, and by many a sad experience, and suffer their teeth to fall out, or to be removed for the supply of artificial ones, while the restoration of many, if not all of them, can be secured for many years, and perhaps for the remainder of life. It is certainly worth the trial; if the gums cannot be induced to advance to their original position, such teeth can be rendered firm and useful by the proper means. They can at least, while their vitality remains, be so much improved as to warrant the endeavor by the Dentist and patient.

I would add that no condition of the teeth so contaminates the breath, as this, except it be a large cavity in a molar or bicuspid with an exposed and sloughing pulp. Here then you have in a word almost the sole cause of a bad breath, as far as the teeth are concerned; and no amount of scrubbing will secure a sweet breath until this cause is removed.

This tartar is composed, in large part, of phosphate of lime, and of course can be dissolved by acids. But these should not be used to remove it, because they also act on the teeth destructively.

" If sloth or negligence the task forbear,
Of making cleanliness a daily care;
If fresh ablution, with the morning sun,
Be quite forborne, or negligently done;
In dark disguise insidious tartar comes,
Incrusts the teeth and irritates the gums,
Till vile deformity usurps the seat,
Where smiles should play and winning graces meet,
And foul disease pollutes the fair domain
Where health and purity should ever reign."

(*Dr. S. Brown*).

CHAPTER XII.

THE FILLING OF TEETH—INCLUDING REMARKS OF GREAT IMPORTANCE.

The subject of this chapter must, from its nature and importance, stand foremost and prominent in all works on the human teeth in their present condition. With the Dentist it is the field of his greatest display of skill in the greatest number and variety of cases which call for his interference. To do this perfectly is the crowning and capital performance of his professional life, the study, and labor, and pride and glory of his handiwork. It has cost him more patient thought and skillful manipulation than any other labor in his whole previous life. Excellence here is to him above titles and emoluments, for it is his crowning honor.

The practice of filling teeth is very old, but it is almost within the present generation that it has been extensively employed, and within a very few years that it has become a positive means of preserving the teeth against decay. These "plague spots," or cavities in the teeth, which have been the principal cause of pain and the destruction of so many myriads of teeth in the present generation and the one immediately preceding, have called to their relief many earnest minds and skillfull hands. The efforts of these men in this direction have been attended with varied success. Many millions of teeth have been filled according to the best of intentions, and many of them have been lost notwithstanding. A vast number however have been rendered useful and comfortable for years, even while this branch of dentistry was young and the operator was comparatively a novice in his profession.

Thousands upon thousands of such teeth have, under the grander developments of this art, been secured almost against the possibility of further destruction at the same point.

As a rule, the durability of a filled tooth depends upon the perfection of the operation—but not always. There are many modifying circumstances present or following the best operations, as well as those less perfect, to change the result. Some teeth are defective in structure and form, and will fail early in spite of the best operations of this kind, unless cared for with great assiduity after being filled. I would by no means dissuade from filling

such teeth. I have seen very many of these con-
stitutionally slender teeth which have been preserved for a great portion of life, by filling, and the
subsequent care taken of them ; neither of which
would alone have preserved them. Such teeth require
more care from both Dentist and owner on account
of their imperfect structure. And this imperfec-
tion, I wish here to remark, appears in families, and
sometimes in localities, and is influenced by the
abundance or scarcity of lime in the food, water
and soil of those localities. Any deficiency in this
respect may be supplied, to a great extent, by
adding to the food and drink some of the salts of
lime. I believe the attempt has recently been
made, with what success I do not know, of increasing
in some way the amount of phosphate of lime in
flour. The idea is a truly philosophical one, and
will doubtless affect favorably the teeth of those
who act upon it.

Teeth which require it should be filled as soon as
cavities make their appearance and have extended
through the enamel. Those between the teeth
will usually be found to have penetrated this flinty
covering about as soon as they can be seen with the
naked eye, and often, by the aid of glasses, before.
These cavities then can be securely stopped ; and

the earlier this is done, the less will be the loss of the substance of the tooth, which is of far more value than the material of the filling, if it were made of diamond.

Whenever cavities are suspected to exist, and even before, resort to the Dentist is proper, whose practised eye will detect them before you have any suspicion of their presence; for it often happens that the first intimation one gets of their existence is from tooth-ache, the " hell of all diseases," as Burns has so strikingly pronounced it. At this time it is often too late for even the most skillful treatment. Do not let the fear that the Dentist will make holes in your teeth, prevent you, for he prefers other employment than making holes in sound teeth.

Do not allow the fear that to begin to call on the Dentist will inaugurate an intimacy with him which you would gladly defer, nor render his services the more necessary afterwards. Your teeth are in far more danger from delay than from the file and cutteau of the Dentist who possesses any skill or honor. The homely adage, "a stitch in time saves nine," is often very true in looking after the teeth, for the presence of caries in one tooth is almost sure to communicate the disease to the contiguous one, which would remain sound if its diseased neighbor was removed or safely filled. This is true of nearly every case where caries attack a tooth on its proximate surface on the side next to its neighbor.

SUBSTANCES FOR FILLING TEETH.

The best known substance for this purpose is pure gold.

First, Because of its malleability and toughness and pliability, by which it can be applied to every part of the cavity so as to make a perfect-fitting plug; and *secondly,* because it is rendered adhesive so that it can be welded together in a solid body, differing in both these respects from all other metals.

Then its purity, and power to resist the action of the fluids of the mouth, renders it the most valuable and durable material to be there worn. Gold is indeed all that can be desired for this purpose,—even its color being scarcely an objection. It is also capable, in skillful hands, of being so packed as to build up and restore portions of the teeth, and even whole crowns of teeth, which have been destroyed. Other substances have been employed for this purpose, but they are all inferior to gold and will hardly bear recommending.

A good serviceable filling must possess the following qualities. It must be made of gold. It must be firmly secured in a clean and suitably formed cavity so as to make a perfect stopping capable of being retained in the tooth, and of excluding all other substances, even air. It must supply all portions of the tooth which are injured or removed by disease; so as to restore the contour and shape of the tooth according to nature. It must be solid as a solid globe or bar of gold, and more smooth and polished. When all the qualities have been secured, it becomes the most skillful and valuable performance of the Dentist, and will both cure and preserve any tooth from any subsequent attack of caries in the same locality *provided nevertheless and always* that proper care be taken of them afterward. For no Dentist will aver that the causes which have produced the disease he has so thoroughly treated, will cease to act upon those teeth or others, by anything he has done, especially if he fails to impress upon you, and you fail to follow this advice—take better care of your teeth than before. Who can tell how much your want of care has contributed to bring you to such grief in the hands of the Dentist; or how much of this

sort of sorrow your subsequent care may save you?
It might be deemed reasonable for one who has
secured in his tooth such a filling as I have described, instantly to reckon it among his jewels!

HOW FIRST-CLASS FILLINGS ARE KNOWN.

First-class fillings will retain a fine finish, and
resist the wear and friction in the natural use of
the teeth, and will preserve the teeth from discolora-
tion, unless it is dead, and will restore it to its
original size and shape. When you have obtained
such a piece of work, you have a good thing and
will enjoy it for the greater portion of life.

GENERAL REMARKS.

Having undertaken this work with a single purpose—the instruction of the family in matters per-
taining to the teeth—and feeling conscious that the
most important branch of the whole subject is this
preservation of the permanent teeth after it has become necessary for you to ask the Dentist's aid, I
wish to be so explicit as to leave no word unwritten
which may contribute to this one object.

Man is confronted at almost every step by emer-
gencies, and is compelled often to cast about for
immediate means to meet them. This is essen-
tially the condition of nearly every individual in our

land, in whose mouth is found any number of the
permanent teeth. In a subsequent chapter I hope
to be able to direct the reader in the production of
sound teeth, and to anticipate those causes of disease and derangement which afflict so many. But
our business now is with the case as we find it.
We find, as a rule, defective teeth in nearly every
mouth and at a very early age. We find them defective in form, size, and color, as well as in position,
distorted often to the degree of hideousness. With
these cases we are compelled to deal as best we can.
In order to do this, we need to look fairly and
truly at the situation, and the means of allevia-
tion at hand. If children's teeth are defective in
size, these children should be especially nourished,
so as to secure perfect calcification, that solidity
and closeness of texture may compensate for
the deficiency in this respect. If they are irregular in position, they should be at a proper age
placed under treatment to correct this defect, because these distortions increase the liability to decay, as well as to impair their usefulness and
beauty. If they are defective in color from any
extraneous deposits on their surfaces, these should
be speedily removed ; and if you are unable to accomplish it, the aid of the Dentist must be sought.

It is better however, vastly better, for you to do this yourself, by a diligent use of the means indicated in a former chapter. I am not writing in the interest of the Dentist in giving these simple, yet powerful means of preserving your teeth. Indeed I have no fear that the Dentist would feel that his special province was invaded, if I should enable you to dispense with his services in keeping your teeth clean, which will also save you many a pang at his hands.

In saying this I am telling you a very important secret, that *perfectly clean teeth* are filled with little or no pain. And there is more in this little secret than the simple statement implies; because the lodgment of particles of food in the cavities caused by caries will immediately form an acid which acts upon the exposed dentine or bone of the tooth under the enamel, rendering it extremely sensitive to the touch of the instrument. Besides, these cavities progress much slower when they are kept clean.

The greatest abuse of the teeth is their disuse. This is emphactially the crying sin in our day with regard to the teeth, and is probably the most prolific cause of their decay. The disuse or misuse of a member or function of the body is by a law of our being tantamount to an abrogation of it. If I bind up my hand or refuse to use its joints, anchylosis or rigidity follows, If man, woman or child feeds upon a watery diet which requires no teeth, by a similar law of nature, the teeth, do what you will, will disappear, and that solidity of structure which insures them against accidents of the commonest kind will be wanting.

Most of the food prepared for our tables is so overwhelmed with gravies and sauces, that teeth are almost superfluous; and nature, revolting against such perversion of her laws, visits their infractions by taking away the gift.*

These reprisals are recognized in all our abuses of the gifts of God, and are witnessed in marked severity on the teeth. For who can estimate the gross amount of suffering endured while an entire set of teeth are being dissolved by caries, or being wrenched one by one from that pearly phalanx so firmly stationed to guard the portals of nutrition —to minister to the perfection of speech and song and to the beauty and grace of every smile ?

The delicious cookery of our own mothers so fresh in all our memories of home, and the delectable mixtures of our professional caterers to the

---

*See Appendix.

palate, cannot afford us the comfort they are designed to bring, because these transient luxuries of the table are enjoyed in violation of the simplest law of physiology—which requires that all food taken into the stomach of man should be thoroughly comminuted in the mouth and mingled with the saliva, to prepare it for digestion.

The practice of soaking the food in a cup of tea or coffee, so common with children, whose facilities for eating are yet unimpaired, robs the teeth of their natural use and insures their early decay. This is an abuse of the teeth which should never be tolerated. Their free use in eating is just as indispensable to their health and safety, as is exercise to the health and vigor to the body. They cannot be preserved without it. So true is this, that we might safely assert that if all solid food taken by man were submitted to the free action of the teeth, these organs would seldom decay. The luxury of eating would also be greatly enhanced, as soon as this habit was established. It should therefore be enjoined upon all young persons as a duty, and ought to be neglected by none.

Munching and nibbling at all hours of the day, when the system has no need of nutriment, is a great abuse of the teeth and of digestion; and it is probably through imparing digestion, that this practice injures the teeth. The effect of this practice may be noticed particularly in pastry cooks, bakers, and confectioners, who almost invariably lose their teeth. A boy put into a grocery, who does not understand the effect of nibbling, on his teeth, and indulges the taste that all boys possess in good degree for the good things around them is pretty sure to spoil his teeth. Such boys often find to their cost that their teeth are going, and yet they are ignorant of the cause.

No boy should go into a retail grocery or candy shop, or bakery, ignorant in this regard; and parents should impress these things upon their sons as they would beseech them to avoid bad company. This subject and the practices here indicated are so prevalent and disastrous to the teeth of this class of young persons, that I ought not to dismiss it without urging upon dentists the importance of pressing this almost sure danger to the teeth upon the persons who are liable to fall into the habit above specified, because of its frequency and sure destruction of the teeth. I know of no personal habit or practice more destructive to the human teeth than this one. It is sure to destroy, sooner or later, the teeth of such persons, despite all that is done for them, except total abstinence from the practice of nibbling of nuts, crackers, cheese, raisins, etc. Not that these articles are in themselves destructive of the teeth or pernicious to health, but the manner and indiscriminate use, with respect to times and seasons, renders them so destructive.

CHAPTER XIII.

Irregularities of the teeth are a source of great trouble, which, if not corrected, becomes very serious. They always more or less disfigure the mouth, amounting often to an ugly deformity.

So conspicuous, often, are the teeth, and so dependent is the beauty of the face on their position and shape, that the straying of these organs from their proper places might seem in many cases almost a worse misfortune than their loss. To suffer them thus to stray, however, is in most cases needless, if not culpable. There is nothing more common than for persons whose teeth are distorted to attribute it directly to negligence on the part of parents; though some are charitable enough to charge it upon their own repugnance and timid aversion to having the infant teeth removed at a proper time.

These distortions are not often the direct result of retaining the first set too long. They are often inherited from one or both of parents, sometimes reaching back a generation or two, and are accompanied by certain forms of face and jaws, being more often seen in thin and narrow faces than in round and plump ones. As a rule, the first or infant teeth should not be extracted to prevent this deformity; it is the surest way to produce it.

The infant teeth are always regular, and, unless distorted by some accident, they occupy their natural position; and, if unimpaired by disease, they serve their time and are removed in a natural and uniform way in regular order. Very soon after one of these is removed, the new tooth hastens forward, and it often happens that two of these occupy as much space as four of the previous ones. Hence the importance of retaining those previous teeth till it is Nature's time to remove them, or until the jaw has attained sufficient size to accomodate the same number of increased proportions, they being almost double the size of the first.

If the teeth are coming irregular from narrowness of the arch of the jaw, this must be expanded —which at a proper age is easily done. The arch of the mouth has generally sufficient extent or surface to accommodate the teeth. It will be seen on examination, that when the sides of the arch are contracted, the center is raised up, as if it were done by force while in a soft state. If the sides were pressed apart, it is evident that room could be gained. The accomplished dentist can correct this difficulty, if the case is submitted to him before it is too late, or before the age of twenty years.

These irregularities are of such frequent occurrence in this country, and so productive of the diseases which invade the teeth and gums, and so repugnant to persons of refined taste, that it is very desirable to call the attention of parents to them. It is next to impossible to keep such teeth clean and free from foreign matter which lodges among them and destroys them; and such teeth are almost sure to fall into decay from their crowded and over-lapped condition.

I would impress more strongly the importance of this subject upon parents, because there is neither necessity nor excuse for irregularities. And where assistance is available, the parent who permits his child to grow up with deformed teeth is justly chargeable with neglecting a duty to his offspring, and will doubtless some day be reproached by his child, for such neglect. Compassion for the child, or distrust of the efficiency of means known to be competent to remedy the difficulty, will not atone for neglect in a case of so much importance. The former is misplaced sympathy, and the latter is ill-timed discretion, for which the child will not be over-grateful in after years when this state of things is irremediable.

Many think all that is desirable to obviate any irregularity is the removal of some obstructing temporary tooth; and many dentists, agreeing with them, forthwith proceed to their extraction. Nothing can be more empirical on the part of the dentist, or injudicious on the part of the parent.

As reasons against this practice it may be urged, that

First, It is contrary to the laws of nature, upon which we can never infringe with impunity.

Second, There is a connection between the temporary and permanent teeth by means of a membraneous cord extending from the neck of the former to the sack of the latter, which must be torn asunder in extracting the temporary teeth. This interferes with the perfect development and deposit of enamel of the new tooth, which process is completed but a short time before the new tooth cuts the gum. Then, by the removal of the temporary teeth, the jaw is liable to contraction, whereas it should expand with the growth of the body and make room for larger teeth.

Mr. Bell mentions the case of a fine healthy boy, with a well developed maxillary arch, being taken to a dentist, who with great dexterity removed eight teeth at once, all of which were firm in their sockets. The consequence was that his mouth so contracted that his new teeth came irregular enough to require four of them to be extracted, in order to remedy the irregularity.

I am pleased to believe that such a dentist would be hard to be found at the present day.

It is also a needless infliction of pain, and should be avoided, as it is always a duty to save a child all the torture possible in the hands of the dentist Many children's teeth suffer great injury through fear of the dentist, because of the terrible shock upon their sensibilities on their first visit to him. They dread him ever after.

None of the temporary teeth should be extracted at random. Let nature alone if she is able to perform her intents, for all her operations are perfect if no casualty makes her deviate from her course; but if any of the causes which I have enumerated obstruct the progress of the permanent teeth, it is right and necessary to interfere. Whenever the following circumstances are found to exist at the period of the second dentition, we may with tolerable certainty predict as to the ultimate regularity of the future teeth :

First, If the parents have regular teeth and have not required the assistance of art.

Second, If the temporary teeth of the child stand a little apart, and especially if they were crowded on their first appearance and have separated, showing an enlargement of the arch of the jaws. In these conditions there is a fair probability that the teeth will be quite regular.

I shall not attempt to describe the means of correcting these irregularities; only to give sufficient information to convince the reader of their importance. The means will be understood by the dentist having the case in charge.

I would advise that the progress of the second dentition be carefully observed; and if these teeth show any obliquity, place the child under the care of a competent dentist, and give him control of the case. His reputation involved will induce him to do the best he can.

There are cases when the extraction of the temporary teeth becomes necessary. When there is a want of consentaneous action between the formation of the permanent tooth and the absorption of the fangs of the temporary one, the latter should be extracted without delay. This will seldom happen except in those cases where the teeth loose the normal vitality, from decay or bruises, which kills them and they become a festering foreign substance in the gums. In this case the prong of the first tooth will not absorb, nor partake in the natural process of development of the new tooth and of its own removal which will require the forceps.

CHAPTER XIV.

TOOTHACHE AND EXTRACTION.

Volumes might be written on this subject without exceeding the experience of every person in this great country. Who is unable to appreciate the recitals of horror and anguish from this source; recitals which are of every-day occurrence, and yet which no pen can describe? One has only to put his hand to his cheek and say toothache, and instantly all present comprehend the nature and location of his suffering, and are ready to offer sympathy or suggest remedies. There is probably no single inheritance of the flesh which distributes its effect so generally among families and individuals as this mortal plague.

The seat of the disorder, the focal point of irritation, is so trivial that a needle's point may cover it, yet so exasperating to the nervous centers, that all the sympathies of the whole system are intensified to fever heat, and the miserable sufferer, almost driven to madness, rushes off to the dentist to "have out" the tooth. The dentist then, often under the pressure of this necessity of the frantic patient. "pulls it out," though it might be rendered useful and quiet for years with proper treatment.

It is often necessary, I admit, to extract a tooth, that being the best thing that can be done; but hundreds and thousands of teeth are being removed under other circumstances, teeth which, with a little patience on the part of the sufferer and with the facilities of this hour, should be treated in a very different manner. Toothache can be cured as readily and safely as any other species of pain without extraction, by means in possession of nearly every dentist, when those teeth are not past usefulness. Myriads of teeth have heretofore been extracted, and more will continue to be, which might have been treated and saved. This is attended with far less suffering than attends the most successful feat of extracting. A little investigation on this point and a little money judiciously invested, will unquestionably save multitudes of beautiful teeth, which, by common consent of doctor and patient, have been devoted to the forceps.

The marvelous facilities for extracting teeth "without pain," so much in vogue at the present

day, are denuding and despoiling the mouth of
vast multitudes of those pearly ornaments which
dentistry is commissioned to save. Oh! the
"slaughter of the innocents" which this mighty
discovery of "anaesthesia" has occasioned! The
dentist is the party most culpable for this wholesale defloration of beauty and robbery of the
human stomach, which is practiced daily for
"tooth-ache," and for the paltry advantage of
adorning the mouth with the questionable works
of art. "A thing of beauty is a joy forever."
What more beautiful ornament of the human
face than a fine clean set of *natural* teeth, to say
nothing of their value in the no less important
province of nutrition? And yet what multitudes
are being deprived, even in youth, of these price-
less gems, under the intoxicating charm of ether,
chloroform and "laughing gas," when, had the
old salutary dread of the forceps ruled in the
premises, they could doubtless have been treated
and saved! The conscientious dentist will firmly
decline to remove any salvable teeth, at whatever
pecuniary sacrifice, though fully conscious that
the patient, believing himself the better judge of
what belongs to him, will go elsewhere and "have
them out." The honest dentist will pocket the

loss of his fee and advice in such a case, feeling better all the while; and the patient mayhap will awaken to a sense of his own loss, when he learns what is daily being done with such teeth without extracting them.

What is said in this chapter relates particularly to toothache from exposed nerves; but there are other aches in the teeth which will be noticed in the chapter on gumboils, alveolar abscess and neuralgia.

CHAPTER XV.

The desire to extract teeth without pain origi-
nated and gave to surgery and the world the most
reliable solace in the hour of fiercest physical
agony which all scientific history furnishes. This
idea, which came like an angel of mercy in benevolent defiance of the terrors of the scalpel and
forceps and to assuage the pangs of parturition,
was the happy thought of a dentist, the late Dr.
Horace Wells, of Hartford, Connecticut. From
that hour to the present, who can estimate a tithe
of the suffering, and apprehension of suffering,
saved to the human body in all civilized lands by
its merciful advent? None can deprecate its
benign mission, nor exaggerate its magic power
over some of the severer vicissitudes of man. All
honor to the memory of him who is accredited
with its discovery, and who died, as it were, in the
very flush of joy at his success—a benefactor of
the race!

But the choicest gifts of man are often abused. The abuse of these agents is in the indiscreet resort to them on trivial occasions, to save a transient pang, and in excessive doses in rash and empirical hands. Most of the deaths from chloroform are from over-doses. The records of our late army, I am told, show only seven or eight fatal cases attributable to chloroform among the thousands who have inhaled it in all sorts of operations for wounds. It is true there may be vast difference in the powers of resistance between the hardy soldier and the delicate lady and child; yet the ratio of fatalities is so small among the many who have inhaled it, that it can hardly be called dangerous in judicious hands. I have never heard of the death of a child by its use.

In speaking thus highly of these agents in formidable and painful operations, I am compelled to utter my earnest protest against an indiscriminate resort to them for the extraction of teeth. *First.* Because myriads of teeth are sacrificed in the frenzied eagerness of persons to be relieved at once and forever from all such sources of pain, not realizing the extent of their loss till it is too late. And, *second.* Because in operations of so frequent occurrence and by all sorts of operators, there is

great difficulty in acquiring sufficient knowledge of
the cases for a prudent diagnosis. I believe that
by far the greatest proportion of fatalities in the
use of anaesthetics has occurred in the extraction
of teeth. There should be no hazard of life for
the simple extraction of an offending tooth. I do
not say that these agents should never be employed
for this purpose; but when it is deemed advisable
to resort to chloroform, at least sufficient time
should be taken and a scientific judgment should
be summoned to the case, or the administration of
it should be declined. *Third.* A selection of the
agent known to be safest should be made. Less
deleterious effects have followed the use of the
nitrous oxide gas, so far as is known; yet not a
few disturbances of the nervous system have resulted even from this, and even death has been
attributed to it, how justly I am unable to state.

Ether, as far as statistics show, is not as liable
to arrest the heart's action as chloroform.

Indeed a distinct and regular pulse at the wrist
or temple must be perceptible during the inhalation
of ether or chloroform to insure safety, because
the shortest possible cessation of the pulse under
such circumstances may be a cessation forever.
Ether or chloroform always depresses the pulse and

respiration, and the cessation of these is death. Nitrous oxide gas quickens the pulse and respiration, and the injuries arising from its use are from overstraining the lungs and heart when these organs are in a weakened state. All these agents are inadmissible to some persons of peculiar, excitable temperaments, and should not be attempted. Persons suspecting themselves to belong to this class should always take counsel before proceeding to inhale them.

One of the constant objections to ether is its disagreeable odor. Chloroform is pleasant to the smell of most persons, and herein exists the danger. It is a subtle agent, and an over-dose may be taken before one is aware of the peril, beguiled by its agreeable odor and intoxicating charms.

But even while I am writing, there is another of this wonderful family of anaesthetics ushered into the world, and this too from the brain of a dentist, Dr. Richardson, who has invented a neat little apparatus, and has suggested, and, I believe, proved, that ether thrown upon a single point in a fine spray continuously, so reduces the temperature of the part in a very short time as to render it insensible, so that teeth may be extracted and other operations be performed without pain or any injurious effects.

Should this be found manageable in extracting teeth, we shall certainly hail it with joy. It is greatly to be desired that something might come into use which may prevent at the same time the pain of severe operations and the dangers of inhalation.

CHAPTER XVI.

ARTIFICIAL TEETH.

Artificial teeth require no definition. What they purport to be is well understood. But the purpose they are designed to serve, and the actual service rendered, are often quite different. I can scarcely speak too highly of them as substitutes for natural teeth which have departed. To those unfortunate ones who have suffered this loss they are indeed a great blessing; and if well made and well adapted, serve the purposes of teeth at least quite respectably. They often appear well, and I have no doubt that many persons derive more comfort from them than they remember to have derived from their predecessors. What might have been done for those lost jewels is not now the question. To the practical point as we find it our attention must now be directed, viz: what are the uses or services rendered by artificial teeth.

No one pretends that they fully supply all the the purposes of a perfect set of Nature's handiwork.

First, They are beautiful, frequently too beautiful for harmony with the proportions and tints of their surroundings. This is a prominent fault of many "false teeth," because proportion, color, form and adaptation are the essential elements of beauty in them, and should all be consulted in procuring an artificial set. A dentist of good taste can judge best how to secure the best effect in this principle of beauty—unless his idea of beauty is fully expressed by "pretty," which is realized if they are small, white and regular, without regard to the size of the mouth, the complexion of the face or lips, the age or sex of the patient. This conformity in size and color to the age, complexion and sex is indispensable to good looks in artificial dentures. In this only can that complete disguise be attained, so essential in all imitations of nature. To secure all these indications of beauty in the construction of these substitutes is the work of an artist—work which long experience does not always produce.

Next to beauty in artificial teeth is the reproduction of distinct enunciation of language. The loss of the teeth impairs not only the beauty of the mouth and face, but also the distinct utterance of speech, to restore which is one of the chief uses of artificial teeth.

Much of this impaired faculty may be re-acquired by practice ; and a perfect adaption of these works of art restores it completely. To secure perfect articulation, these should not be too large and thick, nor too small and contracted. The contraction of the arch of the jaw occasioned by the absence of the roots of the natural teeth is quite considerable, and if these relative proportions are not restored, there will be insufficient room for the tongue, and this embarrassment of the tongue will be detected at once in the speech. The same loss of room will be apparent when the new teeth are too short and arranged too nearly perpendicular, or inclining inwards.

Another consideration to be desired in artificial teeth pertains to mastication. In this particular no definite result can be arrived at. Some persons are much better served in this respect than others can possibly be; and it is no evidence that the new substitutes are not such as they ought to be, if the wearer cannot "eat with them" as well as others. Some persons are more dexterous in acquiring a mastery over this common difficulty than others. Then there is wonderful variety in confirmation of the mouth in different individuals. It hardly seems possible to a casual observer that this unity of design in the common receptacle of food, this common element of beauty, this common aperture whence issue the holy accents of praise, the mellow notes of joy, the grandest utterances of the soul, the wailings of sorrow, and often, far too often, bitterness and cursing, should exhibit such variety in shape, size, and configuration. From this variety arise most of the troubles with artificial teeth. It is due to the dentist that these diffi culties should be fully understood; for he often suffers in reputation when he is in no wise to blame if his structures do not always "fit;" and the reason may be found in this great diversity in the jaws he is required to adorn with his skill. No two mouths are alike, and no set of teeth fitted to one mouth will fit another; and the dentist often suffers pecuniarily by having his best structures left on his hands, a total loss to him, because they cannot, like a shoe or a hat, be sold to another, and he receives for his labor and skill neither money nor thanks. It is well to know always where the difficulty exists, so that justice may be rendered where it is due.

DIFFERENT KINDS OF ARTIFICIAL TEETH.

There is a prevailing aversion to pivoted teeth, or those engrafted upon the roots of the natural teeth; and these roots are often removed when they would sustain an artificial crown for many years. These are the nearest resemblances to nature, of any artificial substitutes, when supplied early after the natural crown is gone, especially in the case of the front teeth or those having but one root.

Of late years these roots have been in most cases taken out, together with many whole and entirely salvable teeth, and whole sets of artificials put in their place. It is the opinion of many experienced dentists that this is an injudicious practice. The haste to substitute the works of art for nature with respect to teeth, while the latter can be rendered comfortable and useful, can only be followed by regret.

If proper care were taken to prepare these roots, they would sustain an artificial crown for many years. But the common practice has been to file them down to the gum, and drill them, and pivot on a crown with wood. This is not the best that can be done in these cases. These roots should be fiiled to the end and supplied with a gold socket which protects the root from decay and renders the whole fixture clean and durable, and preserves in the best possible manner the natural form and expression of the mouth.

CHAPTER XVII.

MATERIALS FOR A BASE OF ARTIFICIAL TEETH.

Various materials have been used upon which to arrange these teeth and adapt them to the mouth. They were first carved from ivory, the teeth and plate being of the same material and of one continued piece. But these were discarded and gold was substituted, and was almost the only substance used for this purpose for many years.

Then a mode was invented by Dr. John Allen, of New York, of making a continuous piece on a base of platinum, which is without doubt the most perfect of anything yet attempted, when the best set of artificial teeth is desired without regard to cost. But it has not come into general use, partly on account of its expensiveness and its greater liability to break. It will probably take the place of all other substances with those able to pay, and, we may hope, will come into general use.

More recently India rubber has been applied to this use, and is more used for this purpose than all other substances. Hard rubber is more used at the present time on account of its cheapness, as it can be afforded at about half the price of gold or platinum. Time only can establish its relative value. It is highly appreciated by those who wear it, and is preferred, as a rule, by those who have worn metal plates, on account of its lightness. But its greatest value as a base for teeth, is that it brings artificial teeth within reach of the many who may require them, who cannot afford to pay for other and superior ones, and will doubtless continue in use unless some valid objection shall appear against it.

But we are not altogether without hope that artificial substitutes themselves may disappear from the mouths of our countrymen. There is certainly hope that, through the wonderful developments in dentistry and the diffusion of knowledge among the people, coming generations will better appreciate and be better able to preserve the natural organs. There seems a prevailing fatality in the minds of the present generation that their teeth must and will decay, and that the one necessity awaits them sooner or later—resort to artificials or depopulated gums for the remainder of life.

I am wholly unwilling to believe that my countrymen, my kindred and mankind are doomed to any such fate. If the present population are degenerated in the size and structure of the teeth, and many are doomed to disappointment in these gifts of nature, I do not believe that God and nature, in punishment of our past abuses, have made our condition unalterable, but that whoever will conform to the simplest laws of physiology may enjoy the fullest use of all physical blessings for the whole period of life.

O ye parents, ye sons and daughters of America, this is the burden of my message to you, the great object of my testimony in these pages, to enlighten you, or at least arouse you to the imperiled condition of these precious, beautiful and indispensable gifts of your Creator, who has kindly *prohibited* you from neglecting and losing them. And I earnestly implore you to reflect a moment before dismissing this subject, and to see what can be done in the household to change the result which we so much deprecate, of losing your teeth, and thus depriving your bodies not only of full nutrition, but of many luxuries which only natural teeth can enable you to enjoy, to say nothing of the pain and suffering incident to their loss.

O ye teachers and educators of youth in physics and morals as well as the faculties of the mind, and who are, with this exception, educators of the entire man in all that concerns his temporal, spiritual and eternal well-being, think; if you can, of one single interest that pertains to the happiness and personal comfort for life of those under your daily charge of so much importance as this, that is not enjoined as a duty and given a place in your text-books as a personal accomplishment and a physiological necessity.

O ye physicians, guardians of the public health and special advisers in domestic hygiene, what do ye for the health and durability of the teeth, to protect them against erosion from contact with some of your necessary prescriptions, and against injury from ichor and sordes during a long and wasting sickness? Sorry work you sometimes make for the dentist, and sorrier still for the patient, after the administration of nitrate of silver iodine and the mineral acids. And you are often boldly charged by the patient with destroying his teeth, while he forgets to give you credit for saving his life. But can you not administer to health without putting the teeth in jeopardy? Excuse me. noble sirs, but I am after the teeth of your patrons and mine, to protect them, if I can, through every accessible channel, so that the whole world may enjoy the natural organs *exclusively.*

O ye writers and lecturers for the public weal, who lash without reserve the public head and conscience! And ye journalists, who see with such lynx-eyed precision all the evils in the world What do ye, besides advertise nostrums and lotions which you know nothing about, to warn or instruct the race which is in daily and hourly danger of grievous disaster in a vital particular? Can you not devote a small corner to an interest which affects every individual in all the land?

I make these special appeals to you, kind friends, because all the interests of humanity are written and lectured to repletion, *save this single one,* which suffers and is behind all others of like and even less importance, for want of illumination. Let there be light shed abroad among the people, equal to that on other subjects of equal interest, and we shall see the mouths of the coming genera-tion adorned with a "flock" of clean white natural teeth, and artificial substitutes will seldom be known except in history, or will be found only in the cabinets of the curious, or among the buried fossils of past generations.

O ye dentists, practitioners of a noble profession, my favorite pursuit for thirty years, in which all the enthusiasm of my nature has been centered, except the little I have bestowed upon our country and the church of the Lord Jesus, come up, I entreat you, come to the rescue—to the rescue and restoration of the teeth of our countrymen, not only nor chiefly to the salvation of those assailed by disease, for the treatment of which you are eminently qualified, but to the recuperation of the productive forces by which teeth are perfectly developed, and to the work of *illumination* as to the means of preservation so little understood by the people. Speak out freely in the public journals and in daily intercourse with patients in your offices, and scatter broadcast, with unselfish liberality, the results of your experience, observation and research, blessing that you may be blessed, giving that you may receive.

Tell the young lady whose delicate appetite shrinks from solid nutritious diet and craves indulgence in things forbidden by the laws of health, that her pale and sickly system is gathering against her, charges, for the persistent infraction of those laws, which she may not be permitted to answer alone, should it ever be her happiness to bestow life and nourishment upon a human being.

Tell the young mother, as she industriously endeavors to seduce her innocent babe into indulgence in the pernicious luxuries of the table, that the best and *only proper* diet for infants is that drawn from the delectable fountains nearest her own heart; that it contains all the elements necessary to develop and nourish in perfect harmony all the parts and functions of the infant life, until the entire set of infant teeth are perfected.

Tell the prospective mother, what every such person may easily understand, that everything she eats, drinks, or feels, affects favorably or otherwise the life and perfection of this new object of her hopes and affections; that she may, by attention to the simplest rules of nutrition, secure to her offspring incalculable blessings for life, not only with respect to the teeth, but to the entire physical being. In all cases of enfeebled nutrient powers and diminished nutrient principles in the daily diet, very charming results can be produced by adding to the food such elements as are wanting. One example of this kind will illustrate this precept and principle completely. It was given to the profession by Dr. Watt, of Ohio. The patient was a lady of slight and fragile form, of weak constitutional powers of body, whose first-born was a daughter, inheriting her mother's physical peculiarities, both mother and daughter losing their teeth at a very early age. The doctor, who was her physician as well as dentist, prescribed for this lady the phosphate of lime in moderate doses, to be taken daily during two subsequent gestations, which resulted in giving her two charming boys, each possessing the finest physical development. The doctor had occasion to watch these boys and note the effect of this early care on their teeth, which was all that could be desired.

Our systems can in nowise appropriate to their structure and sustenance those substances, however necessary, which are wanting in our daily supplies. A former chapter has shown the great deficiency in fine flour, of which most of our bread is made, which deficiency, if not supplied from some source, must perpetually manifest itself in physical imperfections, as shown in the inferior quality and perishable nature of the teeth of many persons.

CHAPTER XVIII.

No pen can adequately portray, nor can tongue
describe the effects upon these organs, which may
be fairly traced to some of the agents standing at
the head of this chapter as their true and proper
causes.

Mercury does not, I believe, act directly on the
teeth; its effect is secondary through the gums
and secretions of the mouth, and when given in
small quantities and continued for a long time it
produces what is called ptyalism. This, however
slight, is hurtful to the teeth, and just in proportion as it affects the juices of the mouth are the
corrosive properties of those fluids increased.
Hence it can be considered only as an indirect
cause of caries. "The relation which the teeth
sustain to their sockets is often very seriously affected and sometimes entirely destroyed by the ex-
hibition of this medicine." Its introduction into
the system is generally followed by increased ac-

tivity of the glands, and in no part of it is this more apparent than in the gums. It sometimes occasions a very rapid loss of substance in these parts, so that the teeth, by the destruction of their sockets, become loose and fall out. Whole sets of sound natural teeth are sometimes destroyed in this way after recovering from an attack of bilious fever. I do not take upon myself, nor is it my duty to assail the use of this agent in the treatment of disease, or to discuss its merits, but only to notice its effects on the teeth. Yet at the same time I cannot forbear to deprecate the profuse and careless manner in which it is frequently employed by persons unacquainted with these subsequent constitutional effects. Its powerful and valuable medicinal properties have gained for it a justly deserved and high reputation. Yet I am persuaded that this popularity has given to its use a license replete with mischief. The imprudent manner in which it is administered during infancy and childhood, while the permanent teeth are being formed, cannot be too strongly censured. A mercurial action in the system at this early period exerts a most deleterious influence on the physical structure of these organs, increasing their liability to decay. Of this I have no doubt, and therefore assert it without fear of contradiction. "The symptoms indicating a mercurial action in the general system are, a slight swelling of the tongue, soreness and increased redness of its edges, soreness, tumefaction and preternatural redness of the gums, with a tendency to bleed on the slight injury, fetor of the breath, vicid saliva in more copious quantities than usual, thickening of the alveolar periosteum, and loosening of the teeth. These may be regarded as criteria of the specific or circumstantial action of this medicine." Some persons are more susceptible to the action of mercurial medicines than others. A single dose will in some instances produce these effects which I have described, while in others several successive doses within a few hours will not in the least affect the secretions of the mouth. The gums, after having been affected with mercury, are ever after more susceptible to the action of irritants, and consequently more liable to become inflamed. "However perfectly," says Mr. Bell, "the effects of mercury may have subsided, even where no permanent injury appears to have been produced in the teeth, and where the common symptoms of its action have entirely ceased to exist, it is not at all unfrequent that after the lapse of several years the teeth become loose, absorption of the gums and alveolar processes take place, and the teeth in consequence are lost."

Indeed, the indiscriminate medication, so common in this country, has undoubtedly contributed vastly to the disorded condition of the digestive organs of our people. The self-prescribed use of medicines which have any power is to be always deprecated. They usually do more harm than good. Those wonderful preparations called "Patent Medicines" are usually compounded by persons who have no knowledge of remedial agents, and are recommended for any and every pain for which they can find a name in some medical book. These names they copy into their circulars, and thus powerful drugs are prescribed in doses suited to their own interests.

Medicines even which are useful and reliable, used without exact reference to adaptability or compatability, or which are not strictly indicated by symptoms, do no sort of good, and of course do harm. All these things are destructive to the teeth, indirectly, if not directly and palpably. A prudent person will never swallow a drug which does not come recommended to him by his equal at least in knowledge. And all those pre-

parations for "cleaning the teeth," which are given
to the public in this irresponsible way are to be
used with extreme caution. They are made to sell.
They look nice, and are very fragrant and have
strange names that in no sense indicate their composition. Some of them possess magic power to
whiten the teeth almost instantly, which is their
worst recommendation, for there is no substance
known to dentists possessed with such power to
whiten the teeth, that does not contain an acid or ·
a caustic alkali. Dentists do not use acids to clean
the the teeth, because they are acquainted with
the composition of the teeth, knowing that they
consist mostly of lime which is an alkali having
a great affinity for acids. These acids immediately
attack the lime in the teeth, and a partial solution
takes place producing that pleasing whiteness at
the expense of the substance of the teeth. The
writer has often met with these whitening agents
being hawked about the country in dram vials, at
the price of one dollar each vial. It is cheap of
course when a single vial will bleach all the teeth
in a neighborhood. But the real character of
these nostrums is seen in the disappointment
which follows in a few days, when the teeth so
magically whitened, blacken worse than before,

and the victim now comes to a sense of his situation, either goes to a. dentist with his calamity or abandons his teeth to the ravages of speedy decay.

The writer has met lately with a fine set of teeth which he has known and admired for their soundness and whiteness. The possessor, a lady of thirty years of age, having a mouth full of hard and firm teeth, came to his office and said, "Doctor, my teeth are all going to pieces!" I looked and found them blackened like a charred forest. I immediately accused her of using iron or nitrate of silver, or some corrosive lotion; but she denied having done any such thing, had not been sick, nor had any sore mouth or throat. I asked if she had used "Sozodont," and she replied she had used nothing else for a year!

Now if any one will show that condition of those teeth *not* to have been directly chargeable to that charming but vicious agent "the Fragrant Sozodont," I will not assail a cosmetic so profitable to its inventor. My observations of its effects I insist have not been at all favorable to its use as a dentifrice. Other eminent and honest dentists have confirmed me in this opinion and contributed to its condemnation. I am of opinion that any dentifrice which is mingled in alcohol is objectionable.

Alcohol is too great a solvent and altogether too fierce an agent to take into the mouth, or to bathe the delicate membrane which lines that cavity. It is poison. Besides, persons not unfrequently acquire pleasure in the flavor of a dentifrice and desire it in the mouth oftener than is necessary to clean the teeth. Those who use snuff as a dentifrice acquire a habit not famous for its cleanliness. So there is great danger that the love of alcohol may insidiously and unsuspectingly establish its fatal delights upon the palate while being innocently employed as a dentifrice.

TOBACCO.

The question is often asked, what effect has tobacco on the teeth. I do not intend to advocate the use of the "weed" as beneficial to the teeth, but shall speak of it as I would of any substance or practice which has anything to do with the teeth or health.

From the observations I have been able to make, tobacco does not cause teeth to decay; it does not seem to affect them in any direct manner, except to wear them out and soil and stain them. Thousands of teeth are prematurely worn out by the practice of chewing tobacco. Yet I am persuaded that very many teeth of slender structure do not come to grief so speedily in persons who chew tobacco, as in those who do not. I do not attribute this however to any virtue in tobacco, but to the mechanical effect of chewing some substance which excites saliva and thus serves to wash away the remains of food from the teeth. Any other inert substance kept in the mouth will have the same effect, a piece of wood or shavings, a quill tooth-pick, some of the resinous gums, etc. I have seen persons who had a habit of chewing their quill tooth-picks, whose teeth were among the cleanest to be met with. Indeed I believe that that to carry a slender piece of quill in the mouth, and to push it through between the teeth often, will most effectually prevent caries from attacking them on their approximal sides.

A nice thin quill tooth-pick should have a place in every *lady's* pocket, as well as gentleman's, and should be plied freely among the teeth after eating. Should a particle of anything which can ferment or sour, gather about the teeth, it should be removed. No single practice would contribute so much in the matter of cleanliness and protection of the teeth, as this, which I might truly say is worth all other processes of cleaning.

ACIDULATED DRINKS—SODA WATER.

I do not recollect to have seen the use of this modern luxury noticed as among the destroying agents with respect to the teeth. And yet I am confident that the free resort to the "soda-fountain" is a resort to a fountain of sorrows to the teeth of the present generation, and will be to coming generations unless attention is called to it.

"Why," says one, "may I not slake my thirst at this sparkling, foaming, delicious fount? Must I be deprived of every luxury at the the hazard of spoiling my teeth; even the simple luxury of bathing my parched throat in hot weather at the delectable shrine whence flows the fragrant nectar of the gods, without meeting at the threshold the unpalatable precepts of some preacher of the *law*, saying 'Drink at your peril'?" Dear mademoiselles and *cher messieurs*, the reason is found in a word. Every time you bathe your fevered throats and tongues with "soda water," so called, you bathe your teeth in free carbonic acid, the most active solvent of your teeth that you could possibly use. This carbonic acid of the soda fountain is made by liberating the carbonic acid from carbonate of lime, one of its constituent parts; and the elements of all substances in nature, thus torn asunder by chemical re-agents, have a marvelous avidity for reunion; hence the rapacious vigor with which this acid attacks the lime in your teeth and devours them.

The acidulated syrups which accompany and sweeten these refreshing draughts are commonly made of mineral acids, and flavored to resemble the fruits by which they are named. The expressed juices of fruits, except of lemons (and this only slightly) do not act so readily on the teeth and can hardly be called objectionable. They are indeed wholesome drinks and grateful to the stomach, and may be used freely in sickness, convalescence, and health, if unaccompanied by carbonic acid gas.

I have known a person who worked during one summer in a manufactory of the "bottled soda-water," to ruin a fine strong set of teeth. He used it freely for drink at all hours, not knowing its effect on the teeth; and his ignorance on the subject cost him a set of natural and also a set of artificial teeth, besides the pain arising from the ruin and loss of the natural ones.

CHAPTER XIX.

THE EFFECTS OF DISEASED TEETH AND GUMS ON THE GENERAL HEALTH.

When we consider the mutual dependance between the teeth and other parts of the body, it does not appear wonderful that, in case these organs are impaired by decay, other parts of the system should suffer a corresponding derangement. It is indeed a law of the animal economy that one organ should sympathize with another; and in fact it often happens that the organ or part sympathetically affected assumes a severer form of disease than the one primarily attacked. With this organ other parts again sympathize, and thus it sometimes proceeds until the whole system becomes implicated in a general and complex disorder.

These organs were designed by an allwise and beneficent Providence for important purposes, and it is necessary to the well being of the body that these purposes should be fulfilled in a perfectly healthy manner.

Mastication appears to be one of the most important functions of the teeth, and claims therefore particular attention in this connection. By this process the food is comminuted and mixed with the saliva, and is thus prepared for those changes which it is destined afterwards to undergo. But when the teeth became incapable of performing this function fairly, aliments are taken into the stomach in an improper state, digestion becomes slow and laborious, and a double duty is thus imposed upon the digestive and assimilative organs, which, of necessity, weakens their powers and hastens their destruction.

It should also be recollected that while food is undergoing trituration in the mouth, it is penetrated by the secretions of the mouth, by which the cohesion of its particles is destroyed and itself is reduced to a pultaceous mass and fitted for the stomach.

Concerning the mastication of aliments and their mixture with the fluids of the mouth, Magendie, an eminent physiologist, proposes the following questions:—Of what use is the trituration of the food and its mixture with the saliva? Is it a simple division, which renders it more fit for the alterations which it undergoes in the stomach; or does it suffer the first degree of assimilation in the mouth? Although Magendie confesses that he cannot satisfactorily answer these inquiries, yet he soon after remarks that mastication and insalivation change the savor and odor of the food; that mastication sufficiently prolonged generally renders digestion more quick and easy; that, on the other hand, people who do not chew their food have often, on this account, slow and painful digestion. He thus admits the importance of mastication, though he cannot tell all the changes it produces in the food. Many physiologists suppose that animalization begins in the mouth; and Magendie himself observes of the saliva, that it is one of the most useful digestive fluids, is favorable to the maceration and division of food, and assists in its deglutition and conversion into chyme. This, however, is a question with which I have little to do in this work, my object being merely to show the importance of the process to healthy nutrition. All must admit that the trituration of food and its mixture with the saliva and mucous secretions of the mouth, are indispensable to perfect digestion. But decayed and loose teeth, and dead roots, turgid and ulcerated gums, and accumulations of salivary calculus often render these liquids not only acrid and irritating to the mouth.

but even nauseating to the stomach. These
fluids, in a healthy state, so far from being injurious to the teeth, are essential to their health
and preservation, but when they become vitiated,
they cause the teeth to decay, they even corrode
the enamel. It has been remarked by several distinguished physiologists, by Richerand and Magendie, that the salivary fluid has a strong affinity
for oxygen, that it readily absorbs it from the air,
and readily imparts it to other bodies, so that
silver, and even gold, are sometimes oxydized by
it. Healthy saliva does not act perceptibly upon
gold. Gold is worn in the mouth for many years
without exhibiting any such effects. In other
cases it has become tarnished in a very short time,
according to the state of the oral secretions.

There are other ways in which diseased teeth
impair the general health. The putrid and offensive
matter that is thrown off from decayed teeth and
from turgid and ulcerated gums, imparts to the
air which passes to and from the lungs a most
disgusting odor, which sometimes contaminates
the atmosphere of a large room and renders it extremely foul for others to breathe. This state of
the breath, although sometimes resulting from
other causes, is a natural and almost inevitable

consequence of decayed teeth and diseased gums, and without doubt frequently occasions serious affections of the lungs.

Dr. Fitch, on this subject, remarks that "nature has formed the lungs most delicate, and sensible of the slightest injurious impressions. She has also finely tempered the atmosphere for its safe and healthy reception into these delicate organs; but accident or disease may render it impure, and cause it, instead of being grateful and invigorating to the lungs, giving health and strength to the whole system, to scatter pestilence and disease through the lungs over the whole body. Nor is it wonderful that the constant inhaling of an impure atmosphere should injure the lungs, since the poisonous matter with which it is charged is brought in contact with them at every inspiration."

But, it may be asked, if a morbid condition of these organs has so great a tendency to impair the health of other parts of the body, why have they not oftener attracted the attention of medical writers with respect to the teeth? It is because the diseases of the teeth are considered as coming more exclusively within the province of the dentist. A few medical writers, it is true, have adverted to their agency in the production of other diseases, and the correctness of their observations must carry with them conviction of their truth.

Dr. Rush observes, " When we consider how often the teeth when decayed are exposed to irritation from hot and cold drinks and aliments, and from pressure and cold air, and how intricate is the connection of the mouth with the whole system, I am disposed to believe that they are often the unsuspected causes of general and particularly of nervous diseases." " When we add to the list of these diseases the morbid effects of the acrid and putrid matters which are sometimes discharged from carious teeth, and from ulcerated gums created by them, also the influence which both have in preventing mastication, and the connection of that animal function with good health, I cannot help thinking that our success in treating all chronic diseases would be very much promoted by directing our inquiries into the state of the teeth in sick people, and by advising their removal in every case when they cannot be retained without irritation."

The influence of diseased teeth on the stomach and other parts of the system was noticed by Baglivi as early as the beginning of the eighteenth century. He remarks:—" Persons whose teeth are in an unclean and viscid state, though washed daily, have uniformly weak stomachs, bad digestion, an offensive breath, headache after meals, and generally bad health and low spirits. If engaged in business or study, they are impatient and are often seized with dizziness." Physicians of the present day are beginning to pay more attention to this subject than formerly. They have discovered that many local and constitutional affections of the body are produced by an unhealthy condition of the teeth, and that they cannot be remedied without first restoring these organs to health.

Many eminent physicians have confessed the embarrassment which they experience in treating chronic and nervous affections of patients who have diseased teeth, and have directed their patients first to the dentist, whose attentions to the teeth alone have often been sufficient to restore the health. If it is true, as a late medical writer observes, that "the mainspring in the cure of disease is the subduction of its causes," and if diseases of the teeth and gums exert a morbid influence on other parts of the body, then it is essential that these should be perfectly understood, and that in the treatment of disorders produced by them, such means should be employed as would most effectually remove them.

Having thus briefly adverted to the effect of dental diseases on the system, let me present a few facts in support of this doctrine, as they were collated by Prof. Harris:

First. "A lady of high respectability was advised by her physician to consult Dr. H. in regard to her teeth. She had been in delicate and precarious health for six years. She had taken much medicine and had visited watering places, without relief. Her stomach was so much deranged that the lightest kind of food distressed her for a long time after it was taken. Her whole nervous system was so disordered that the quick slamming of a door or any other sudden noise, would almost throw her into convulsions. Her eyesight was much impaired, her head was affected with an almost constant swimming or dizziness. On examining her mouth, I found a large number of her teeth involved in extensive caries, the gums tumid, soft and spongy, and ulcerated along the edges. This lady's teeth were cared for, her mouth put in a healthy condition, and her health was completely restored. Furthermore, she had enjoyed good health till her teeth became diseased. I think it cannot be doubted that her illness was occasioned by this morbid condition of her teeth."

I might multiply these cases almost indefinitely in which diseases have baffled the efforts of the best physicians, but yielded readily to treatment after attention had been paid to the teeth, so that irritation from this source no longer existed. Even well defined symptoms of pulmonary consumption have been arrested by restoring the constitutional balances disturbed by diseased teeth and gums.

Dr. Rush mentions two cases—one of epilepsy, the other of rheumatism—which, he says, were produced by decayed teeth. He directed them to be removed, and he adds, " I had the pleasure of seeing both cases perfectly cured."

Earache is often caused by a bad tooth. A dentist of my acquaintance told me that he was entirely relieved of a rheumatic affection in his shoulder by having a troublesome tooth extracted.

Hufeland enumerates firm and sound teeth among the signs of long life. For good digestion good teeth are extremely necessary; and one may therefore consider them among the properties requisite to long life, and that in two points of view. First. Good and strong teeth are always a sign of sound and strong constitution. Those who lose their teeth early have in a measure taken possession of the other world with a part of their bodies. Secondly. The teeth are a great help to digestion, and consequently to restoration from sickness. Many troublesome abscesses about the face and jaws have their origin in diseased teeth. One case came under my own observation, of osteo-sarcoma in the cheek, resulting in death, which had its origin from this source. It often happens in a diseased and ulcerated condition of the lower wisdom teeth, that as soon as they become swollen, it is almost impossible to get the mouth open so as to have them removed. The consequence is, that abscesses are formed which break through the jaw at the angle, and discharge until removed. leaving an unpleasant scar on the outside. I have known a long sickness to result from one of these, because of the extreme difficulty of removing the teeth.

These wisdom teeth should never be suffered to remain in the mouth after their usefulness is gone. It is sometimes necessary to remove them when sound, on account of their partial concealment under the gums, causing inflammations, which often extend to the fauces and throat.

Many ladies suffer intensely from their teeth, particularly during the period of gestation, and often during the whole of that period. And very frequently they are assured that it is one of the incidents of their condition, and one which they must endure because of the hazards of doing anything to their teeth under such circumstances. But *I* can also *assure* them that it is by far the best time they will have for many months to come. Any time, as a rule, before parturition is better for necessary attentions of this kind than for several months afterward; because the general health of most women is better then than it may be for a considerable time after that event. And it is not profitable for either mother or child to suffer from such local irritation that can be safely and easily relieved. I am fully justified in giving this opinion from long experience and numerous cases of the kind which have consulted me at such times.

"It has been established by an English surgeon (Dr. J. P. II. Brown,) that this condition of the teeth commences to manifest itself in many women after fecundation, for during that state the mother is called on to furnish precisely this calcareous element for the development of the foetal structure; and when this assimilating material is insufficient, the foetal organization must be supplied at the cost of the bone proper of the teeth—a material difficult of repair, but admitting of ready and rapid deterioration. Dr. Magitot, in his new treatise on dental caries, demonstrates that the saliva of the mouth, in consequence of the change in its composition in various diseases, as typhoid fever, dyspepsia, etc., exercises a hurtful influence and is a true cause of dental caries."

"Experience," says a common proverb, "is the best schoolmaster," and it may in the main be true. But in the things of which I speak, I would not leave the learner of to-day to experience individually the regrets of his predecessors. Regrets the most painful are, in the matters of which I write, unavailing and profitless to the sufferer. It was a remark of Henry Ward Beecher, that if his foresight was as good as his "hindsight," he might avoid many difficulties. This retrospect, in order to be sanitary with respect to the teeth, must of course include the mistakes and follies of others. The point I wish to make is, that if mothers and farthers, and indeed sisters and brothers, whose experiences have been regretful, would set a proper example, the younger members and successors might take note and avoid much suffering.

The effect of example on the part of mothers, the silent example of cleaning and watching their teeth daily in presence of their children, and their expressed appreciation of clean and beautiful teeth as a personal ornament and necessity, and of the infinite value of sound natural teeth as compared with artificials, will accomplish more in this direction than many precepts without example.

The experience of all dentists will confirm this doctrine and justify the prominence I have given it; and there is no dentist who will not gratefully acknowledge the assistance of a mother's example and co-operation in his honest endeavors to preserve her children's teeth, applying to him for assistance. The power of this example and co-operation does not end with childhood and youth, but almost invariably establishes and confirms a habit of life-long continuance and reward.

But what shall I say of the reverse side of this subject? It need not be named nor characterized. Inference completes the argument and applies it.

CHAPTER XX.

THE DUTIES OF DENTISTS.

One of the subjects which has occupied my thoughts in view of the duties and responsibilities of every man of us who is engaged in the teaching and practice of our young and thrifty profession, is the generally abnormal condition of the teeth of Americans. To this condition of the teeth of our countrymen may be justly traceable a large portion of physical suffering for which our own and kindred professions are called to prescribe. To consider the causes of this anomalous departure from a healthy standard, to search out and designate remedies which may control these effects, to save the human teeth thus diseased, and to contribute in some way to the restoration of those disordered functions of the body which result in the destruction of these important organs—these are the objects, in a beneficient sense, of all our labors, discussions, and researches, and these surely furnish cause why a man should rejoice to be a dentist. Here lies the nobility, the skill, the dignity, the greatness, and growth of our profession. To ascertain the causes of these astonishing effects, that we may not only in some degree arrest them, but be able to turn back the tide of disintegration and wholesale dissolution which threatens alike the durability and safety of the teeth of the American people, is a labor of love which may well challenge our individual and united energies and engage our hearts.

Where shall we begin and what shall we do? These are the questions of the hour, and momentous questions they are. I would for myself reply, *at the beginning*; that is, we must take cognizance of the teeth of the children who are to be born. Here is the beginning of that catalogue of sorrows to which our profession is called to administer. It is at the germinal point and in the symptomal stages that prophylaxis or preventive treatment is indicated. Active or aggressive treatment afterwards may do something to arrest or change a vitiated action in the organs and tissues of the body. But he who would wisely and prudently anticipate results which he may not be able to control, will see to it that the deranging processes are nipped in the bud. The axe must be laid at the root of every vile tree, if you would destroy it root and branch. All reform must be radical, if evils are to be removed.

So if our profession shall contemplate the restoration and salvation of the teeth of our countrymen, we must search for the causes of their diseases somewhere near to the beginning of these effects. These causes I consider to be congenital in their origin, if not primogenital. They are not beyond the reach of remedies, but are easily and safely controlled by prophylactic treatment. If it is true in any sense that the ounce of prevention is worth the pound of cure, why not in the special province of the dentist? If then we are to direct our diagnosis and apply our prophylaxis at the germinal seat of this difficulty—and nobody questions this philosophy—we must aim at the enlightenment and secure the co-operation of *mothers.* The maternal heart yearns with irrepressible anxiety for the safety and well being of her unborn offspring. All the sympathies of her nature are intensified from the first conscious moment of so interesting a prospect. She is anxious to know what she may or may not do to affect favorably or otherwise the life or perfection of her new treasure. If the physician does not advise nor caution her with respect to the teeth of this germinal man or woman, and the common nurse cannot, the dentist must. The teeth are our specialty, and whatever affects these organs adversely or otherwise, especially concerns us, whether in the foetal state or infantile, whether in childhood, youth, or manhood. The question then comes, is it practicable? I tell you far less sensible theories than this have been proved practicable, which have been scouted by this great wise world as illusory, but which a later and wiser generation have adopted as substantial and true. How recently the theory of the gifted Hannaman, "similia similibus curanter." was greeted with the execration and malediction of every medicine man in Europe and America! Yet upon it is based a system of treatment which will compare favorably with any other of the same age. and development in either hemisphere.

Eminently true is this of many of the most beneficent improvements of modern times. That which is denounced as theory and humbug to-day, becomes often a reality to-morrow and a demonstrated truth. To look wise and wag the head and roll the tongue, is far easier than to investigate and *prove* a proposition to be false. Galileo, Columbus and Luther barely escaped martyrdom for boldly asserting and maintaining their convictions in the midst of surrounding superstition. skepticism and stupidity. And yet those very doctrines are matters of course with us to-day, because unquestionably true. The same might be said of nearly all the great reformers and benefactors of mankind whom God has raised up and held as beacon lights along the perilous track of beneficent history. The popular cry of humbug, and the scathing artillery of ridicule, and the blighting merciless incubus of poverty, have driven many a modest but gifted genius into retirement, neutralized or abridged his opportunities, and robbed the world of the fruit of his labors. But, thanks be to God, truth, though crushed to earth, is elastic as it is eternal, and will rise again and assert its supremacy over all embarrassments, and flourish with an omnipotence that is divine.

The prophylaxis I would suggest to secure this object is dietetic rather than medicinal. Good wholesome food in suitable quantities, selected not according to the demands of a vitiated palate, but with a rigid adherence to strict physiological principles, is the best protective of teeth. The milk and cream and butter of your "Alderneys" are far better than sweetmeats and pastries, to make bone and muscle and teeth.

This is the first step in our return to the old paths from which young America has so blindly

and disastrously departed with respect to the teeth.
The universal need of a dentist at every cross-road
in the country and on every square in the city, is
proverbially American. Teeth in the older communities of Europe and the East are fifty or a
hundred per cent. better than here, except in the
large cities, where, as here, culinary civilization
has outrun healthful progress and fallen into re-
prehensible barbarism. The delectable cuisine of
most of our well-to-do families is hurrying into
oblivion more of otherwise good natural teeth
annually than S. S. White and a host of others can
make of their best imitations. Most of·the food
prepared for our tables is so overwhelmed with
gravies and sauces that we have only to suck to
get our supplies, and our boys and girls scarcely
mistrust they are weaned! The disuse or misuse
of a member or function of our bodies is, by a law
of our being, tantamount to an abrogation of the
member or function altogether. If I bind up my
hand, or refuse to use its joints, anchylosis follows.
If your children feed upon slops which require no
teeth, by a similar law of nature, *their teeth will
disappear, do what you will.* Answer me not that
this reform is impracticable. We can gird up the
loins of our understanding and go forth to the

field of duty and renown, scattering broadcast with unselfish liberality the results of our experience and research. Oh! that my inspirations were proportioned to the magnitude of this subject, and my tongue as the pen of a ready writer, and my voice as the diameter of the earth multiplied into its circumference, that I might give to these things the demonstration they deserve.

Shall we, dentists, be content, though experts in our art, merely to patch up and repair the waste places in the mouths of the few within our reach, while wholesale desolation and ruin threatens alike the teeth of the many present and to come. Nay, rather let us speak out with voice and pen in convention and in the public prints, and in daily intercourse with our patients, sounding the notes of alarm and proffering wholesome counsel which shall be like "apples of gold in pictures of silver." Tell the grocer's boy as he nibbles and munches, hour by hour, the things he is hired to sell; tell the confectioner who lives that teeth may be destroyed; tell the nibbler in mother's pantry; tell all the intermeal munching community of candy eaters, and the devourers of pies, pastries and slops, that such practices will speedily and inevitably destroy their teeth, and bring them to grief in the hands of the dentist. Tell the young lady whose fastiduous and mincing appetite shrinks from solid nutritious diet and craves indulgence in things forbidden by the laws of health, that her pale and sickly system is gathering charges against her for the persistent infraction of these laws which she may not be permitted to answer for alone. Tell the young mother, as she industriously tries to seduce her innocent babe into indulgence in the pernicious luxuries of the table, that the best and only proper diet for infants is that drawn from the delectable fountain nearest her own heart. Explain to her this all-important physiological truth, that she must take into her system, in her food or otherwise, a sufficient amount of lime salts, as well as farina and *fibrine,* if she would impart health and vigor to her offspring with respect to the muscles, bones and teeth. I do not insist that she shall eat lime phosphates in any considerable quantities, but that she should diet largely on such substances as contain in unabated proportions these qualities. Hens are not required to live on calcined bones and clam shells, to save them the mortification of shell-less eggs, if they can only get plenty of whole grain, which contains of these qualities 85 parts in 500, while fine flour contains only 30.

The poet Barlow sang this doctrine in revolutionary times a hundred years ago, in his humorous poem on Hasty Pudding:

My father loved thee through his length of days,
For thee his fields were shaded o'er with maize;
From thee what health what vigor he possessed,
Ten sturdy freemen from his loins attest.
Thy constellation ruled my natal morn,
And all my bones and teeth were made of Indian corn.

Where is the man whose teeth in youth were familiar with the crackle of roast potatoes, brown bread and Johnny-cake, and whose body was early nourished with beans, buttermilk and "bonny-clopper," who cannot enjoy to-day, without artificial aid, those invigorating viands of his boyhood with a relish and satisfaction second only to the sacred recollections he bears of his mother?

I would not hesitate to prescribe these substances in a free state in suitable quantities, if I had any suspicion that the daily food did not contain them in requisite proportions. Nor would I heed at all the fears of any one whose captious cautions had so muddled his judgment as to cause him to apprehend embarrassments of a monstrous kind. I would not deny to dame Nature the right or power to make her own selections as to quantities, if the qualities are only placed within her reach It is the want of these qualities in our daily food that causes all this trouble of which I speak.

I have often thought that these dissolving processes going on in the teeth, this breaking up of such solid continuity, was nothing more nor less than a furious spasmodic effort of the system to obtain for immediate use *the very elements of which teeth are composed.* And thus they are dissolved and taken into the stomach to supply in part this demand.

The fluids of the mouth and stomach, when that healthy equipoise between the acids and alkaloids is destroyed, become solvents of such potency that they will almost dissolve the mucous membranes themselves, and would if these were made of lime. This may be theory, and may look a little airy to-day, but you may shout amen to its truthfulness to-morrow. We now only know in part, and discern truth through a glass darkly. But the day is fast approaching when we shall see in their true light and utility, many things we do not now consider; and among them will be this very subject about which I am so earnest and didactic.

Great and important as this reform is, it is not so formidable nor so difficult of accomplishment as many with which we all are familiar. That a reformation in this direction is desirable none will deny. How and by whom shall it be accomplished? These are the questions that I would make special with every dentist. We have seen, if I rightly apprehend, the primal cause of this disorder and its remedy, that a better state of things in this regard is attainable, and to attain it is our legitimate work.

None can better appreciate than the dentist the value of a good set of teeth, nor estimate their loss more accurately. None can better understand their diseases, nor detect so early and so minutely the destroyer's stealthy footsteps, nor discern so remotely the signs of his coming.

But the cases with which we have had most to do are by no means symptomal; they are mostly advanced stages of disease, or stages past recovery. Our profession has made great progress in the treatment of teeth, even in these calamitous conditions. But to anticipate and prevent has not, it seems to me, sufficiently occupied our attention.

Recuperation, the recovery of the lost estate of the teeth of our countrymen, is one of the grandest enterprises open to the dentist, and one which will abundantly reward all right-minded and earnest men who seek the advancement of our calling to the dignity of a learned profession.

Let us come down to first principles and direct our inquiries in the line of primal causes, if we would learn how to control resulting effects. Let us pursue our calling with zeal, looking for fruition and reward to the future.

O happy day! O auspicious time to come! Make haste and tarry not on the confines of any past or present attainments, till dentistry, youngest fledging from the groves of the academy, shall rise, with healing in her wings, for these special woes of my kindred, my countrymen and mankind.

CHAPTER XXI

There is one fruitful cause of decay of the teeth, not noticed in the chapter on that subject, of which I ought to speak. It is one of which no mention is made, so far as I remember, by any writer on the diseases of the teeth. It is *the unnatural habit of breathing through the mouth.* The teeth and mucus membrane of the mouth are provided with a peculiar fluid, which is ever present, in a healthy state, in proper quantity to keep those organs in a moist condition, which is as essential to their healthy condition as is the tear drop to the eye, or the senovial fluid to the joints.

Breathing through the mouth not only drys this natural moisture, but it does a great deal more, it actually keeps the teeth bathed at every expiration with a current of carbonic acid of sufficient potency to attack the teeth. Especially during the hours of ·sleep, the mouth being open and the lips are in a measure retracted, the teeth become

dry and are exposed to the constant current of air
which is expelled from the lungs more or less
charged with the great destroyer of the teeth—
acid, and then the saliva which remains and is retained between the teeth absorbs this acid at each
successive expiration until it acquires sufficient
strength to etch the enamel and dentine until
cavities are formed between the teeth. I think all
observers will agree that those teeth which are
thus exposed decay much sooner than those habitu-
ally covered by the lips, and are kept moist by
this natural and protecting bath, good saliva.

The natural condition of saliva is slightly alkaline, and departures from it in the varying moods
of health are always acid, and the least preponder-
ance of acid in the composition of the saliva at
once becomes destructive to the teeth. Either of
these conditions are readily detected by those simple tests employed by the chemist to ascertain the
presence of acids or alkalis. A small bit of litmus
paper moistened in the saliva present in the
mouth, especially between the teeth, will instantly
turn a bluish purple when the least acid is present.
The use of a dentrifice containing soap chalk or
soda will remove this acid condition of the saliva
for the time being, but perfect digestion is the only

cure. Indeed it is not too much to affirm that the very fact of breathing into the mouth from the lungs of air so charged with acid, as is all expired air, is sufficient of itself to keep the saliva that may be present in the mouth in an acid condition. It requires but a slight trace of acid to disturb that healthy equipoise between this slightly alkaline condition of normal saliva, and an acid condition so damaging to the teeth. The sour taste in the mouth during an attack of indigestion may often be traced to this cause alone, expelling through the mouth the vitiated air from the lungs. It is true that the teeth are covered by a hard flinty coat called enamel; but this is no protection against decay in the teeth. It is not designed as a protection againt decay, but against wear. Observe how soon your teeth change their shape and diminish in length as soon as the enamel is worn off or otherwise denuded from the ends of the teeth. That fabulous grandfather who "*had double teeth all around,*" acquired the appearance of such remarkable teeth by their long continuance and service, and by the peculiar manner of their occlusion. It is also true that those of our ancestors who retained their teeth to old age did not leave them out nights exposed to the cast out breath.

which is not even fit to breathe again, and cannot be with safety until it has imparted its impurity to the vegetable world, which thrives by the effete products of animal life. If one should sleep for a single night even with their eyes open, he would readily discover the use and advantages of eyelids.

APPENDIX.

NATURE'S LAW OF REPRISAL.

I have stated in the course of this work that the greatest abuse of the teeth is their *disuse ;* and I there remarked that this result was in accordance with a law of nature, that the neglect or disuse of any member becomes in the end tantamount to its abrogation. The effects of this law of reprisal are so manifest in our day, that one reflection forces itself upon me just here, not contemplated in the text.

It is that *the sacred feeling of maternal love* is endangered by it. Witness the appalling increase of foundlings, of infanticides, of the reckless and criminal interference and murderous outrages against fœtal life, committed in these days. What true maternal heart is not shocked at these daily reports in the papers? Our country is flooded with advertised nostrums whose *murderous* aim is directly at the very embryo existence of human beings.

Any interference or interruption of the process of gestation is murder, and should be characterized and treated as such, because it is a blow aimed at the fountain and source of society, and of national life and families as well. If individuals are extinguished, so may families and nations be. Populations and nations are the growth and additions of individuals to the existing numbers; and if the number of births is not in excess of the deaths, populations must of course diminish and ultimately disappear. And if nations and peoples whose opportunities and privileges have been exalted by culture and religion, transgress the first principles of life, surely *they must be supplanted*, and their places be filled by the simpler children, whether they are savages or wise and God-fearing men.

The history of nations is a continued series of examples of this law of reprisals, even to *extinction*, for physical and moral transgressions alike, since these go hand in hand. To expect high moral culture and purity in a depraved body is to be disappointed. Sterility is the inevitable consequence of all voluntary unfruitfulness. The number as well as the vigor of individual members in the best priviledged families in our land is daily diminishing; while the increase of our population, from its own resources, comes in large measure from the lowly. And then, too, comparatively few of the children born to luxury, or even to abundance, live to maturity, because excesses and indulgences seem to follow in the ratio of the facilities to gratify them, leading to and ending in perversions and violations of physiological laws, and the suffering of their penalties; while those little multitudes and nascent communities which surround the table of the humble cottager remain often unbroken, and their vigorous groups, reared in simple dietic habits often, I am happy to know, furnish the best specimens of the race.

But it is when the well populated homes of these simple children of nature become also the homes of the purities and sanctities of life, the homes of virtue, and truth, and holiness, that such fruits are seen. Squalor and vice yield no such results. The extremes of poverty and luxury demonstrate alike the results of violated law.

But it is neither from abundance nor want, as such, that dangers spring. It is from wrong views and practices incident to each of these. In the one case abundance without philosophy leads to surfeit; and in the other, the desire to attain, without philosophy, leads to over-exertion. Both conduce to physical, moral, and intellectual effeminacy, and hence to extinction.

The old New England families of from seven to half a score, reared in simple Puritan habits and nurtured in sterling Puritan virtues, gave laws and character to the nation, and leavened its entire civilization with the true principles of progress and patriotism. But those to whom this honor is due are fast passing away; and it remains a question of much significance whether the present and coming generations will hold the power in preserving and perpetuating the institutions and the social, religious and judicial foundations established by their fathers, against the influx of emigration, and other decimating processes apparent in this country. Statistics show than the average of children born in native New England families is only three and a fraction; while in the families of the foreign population it is double or more. All this retrogression in our American families is the work of a very few years. If this is true, how long will the final catastrophe of extinction of the Anglo-American race be delayed, and what will be its effect upon America and the world; or rather what will not be the dreadful result?

It is said that in one considerable district in a city at the West, not one child of American parentage has been born in three years. It is not uncommon to see parents leading by the hand, or drawing in a little wagon, a solitary child —and that a sickly waif—its very countenance and figure foreshadowing the brevity of its existence. And they are to have no more; this is the first and the last, and soon it is gone, and with it, possibly, the very instinct of ·maternal love revoked for want of occupation, dead for lack of sustenance. Or it may be that a little latent vitality lingers from a once vigorous force, to be expended upon pet brutes, and their house is left unto them desolate, emptied of its ornament and its life. It may be an estate with family associations and history, to pass to other names for lack of family successors. Not so with the great mass of our imported population whose peculiar progeny are filling every nook and corner in the land and are threatening to outnumber all other populations. Does this look like preserving and perpetuating our institutions and manners, our civilization and nationality, our literature and free Gospel, our position and prestige among the nations?

The older nations of Europe are improving in the quality of their populations, sending hither their excrescences. And if it is our destiny to school and purify the *dross* of the world, we must preserve enough of the original stock at least for specimens and examples.

These two forces, the sterility of Anglo-Americans and the unabated fruitfulness of the imported population, are the threatening horoscope of this nation, and foreshadow a future which no true American can pleasantly contemplate. It is one, however, which every American of either sex should wisely and seriously consider—especially those who contemplate marriage.

There are other aspects of this subject beside those here indicated, which are equally important, but which more properly pertain to the province of the moral teacher than to one who is endeavoring to exhibit it in a national and physiological point of view, and in its desolating effect on the central charm of woman's character—maternal love. Woman, denuded of this potent element in the very seat and center of her affections, is no longer the woman of history, of poetry and song, nor can she remain the central figure in the family group and in the heart of man. Her position of equality by the side of man, even her dominion in his heart, over his ambitions, his endeavors of all earthly kinds, must change. The oneness of the twain, the unity of all the temporalities of two individuals for life, may find its point of divergence here. Instead of remaining the peerless queen of man's highest earthly aspirations, she dwindles to his toy, his plaything, or worse. Woman was made to hold empire in the heart of man, and to wield a mild and gracious sceptre. This she does by four cardinal powers of her being—as a daughter, a sister, a wife, and a mother. These are points of mighty influence, from which radiate, as from distributing centers, her powers over the world. Under these enchanting names, she exerts all her transforming influence on human destiny. O then, ye daughters of Christianized America, look, I beseech you, at what is passing daily before your eyes, and see if your hearts do not freshen to a more healthful sense of your responsibilities and duties as daughters, wives and mothers, in an age which demands the best endeavors of the best of women as well as the best of men. This is the serious conviction of every philanthropist and person of reflection in the land, reiterated and urged in the interest of humanity by an earnest lover of the race, and under the pressure of an abiding sense of duty that springs from the very nature and imperativeness of the subject.

I write in extreme diffidence, without name or prestige as a writer or moral critic, with the simple hope that these pages may be read by some one for whom they are written, and in the earnest wish that the thoughts suggested may impress the judgments and arouse the consciences of the honored and privileged but imperilled daughters of my country. This is an age of easy virtue, of fascile dereliction from duty, of most captivating temptations to shrink from the wholesome restraints of divine law. Whoever listens to the enchanting but delusive notes of the destroyer is in much greater peril even than I have shown. Not only will the heart of woman be deflowered of one of her distinguishing charms, but her moral and physical natures will also be in extreme jeopardy from the energy and absoluteness of the law of reprisals.

Of her moral dangers, my present purpose does not lead me to speak. I am dealing with the physical and tangible —not with the ethical and spiritual nature, though that also is amenable to this law and must report to its judgment seat *in the life that now is.* "By grace are ye saved," is a truth without comfort in the physical life. Here there is no provision for grace. Whoso sins, forfeits. The sun which reveals beauty and proportion in forms and colors, shines in vain for the fair lady who immures herself in darkened dwellings, or hides her alabaster face in a thick

veil in the street. She defeats the object of all her solici-
tude. Her complexion can never be beautiful; because
light—sunlight—is as essential in the production of beauty
as it is in revealing it. I do not intend to aver that a lady's
face must be browned by the rays of the sun to be beautiful; but that the light of the king of day must be permitted
to contribute, as he invariably does and always has, to all
beauties of tint. From his long experience and known
power, he can do it better and more satisfactorily than all
the cosmeticians of Paris, or the world, who cater to the
shrine of beauty. The exclusion of light from our dwell-
ing is, in the opinion of eminent physiologists and pathol-
ogists, one of the chief causes of muscular and nervous
debility in American females. Light is as essential to the
blood and skin and eyes, as air is to the lungs, or food to
the stomach, if it is desired to preserve health and vigor
and consequently beauty. So, too, if one half of the body
be compressed by external trappings, this restrained portion
will be shriveled and distorted, while the other will exceed
its destined size, and that exact equipoise of weight and figure which give symmetry of form and grace of motion so
much admired, whether in standing, walking, or dancing,
and in which woman by universal consent excels,—will be
wanting. No amount of practice or training can compensate for this want of balance and proportion and perfection
of configuration as seen in a perfectly graceful woman.

I am admonished that brevity, which is designed to rule
in this work, forbids amplification; but by no means does
the nature or importance of the theme, nor inclination,
drive me from a field so full of interest to every thoughtful

mind of patriotic and humanitarian impulses. Americans will read paragraphs, listen to short sermons, think a moment, and be off after the new and often the dangerous. Didactic lessons are too often ruled out. Satire and panegyric fail to convict the quick, active and self-reliant American mind. Labor and logic, supported by careful statistics, may also fail to correct or arrest the rushing torrent of American follies

But the example of the good and wise can do much. Too much preaching and hearing, fosters stupidity; but the practice of virtue saves from vice. My countrymen are intelligent, and a word to the wise is sufficient. But the giddy and thoughtless will be indifferent still.

TESTIMONIALS.

• • • --- -

NEW HAVEN, June 7, 1869.

My acquaintance with Dr. Woolworth's skill and enthusiasm in his profession, makes me confident that his little book entitled "Our Children's Teeth," ought to be published, and will well repay the publisher. The book is so directly practical in its aim, and is so characterized by common sense in the author appealing to common sense in his readers, that I cannot but think it likely to obtain a wide circulation.

LEONARD BACON, D. D.

I have been for many years familiarly acquainted with Dr. Isaac Woolworth, of this city. He is a thoroughly educated physician, though he has devoted himself mainly to the profession of a dentist. Of his consummate skill in this department, I can speak from much experience, and with the fullest confidence. Dr. Woolworth has prepared a work, designed for popular use, on the diseases and dangers of the teeth, and the best means of guarding against them. I have conversed with him on the plan of the work, and have heard portions of it. It is my belief that, if published, the book will be favorably received, and will be found widely and greatly useful.

JAMES HADLEY,
Professor of Greek.

Yale College, May 17, 1869.

YALE COLLEGE, May 31, 1869.

I have made myself acquainted with the plan of Dr.
Woolworth's work on Dentistry for the Household, and
have been struck with the propriety and fullness in the selection of topics, and their practical bearing. Such a work,
giving instructions about teeth from the early times of childhood, strikes me as greatly needed, and is likely to prevent
much loss and much suffering.

THEODORE D. WOOLSEY,
Pres't of Yale College.

During an imprisonment of several hours in the dentist's
office of Dr. Isaac Woolworth, of this city, I had an opportunity some months since to look over a manuscript which
he had prepared on the proper care of the teeth, and especially the teeth of children; and I thought, and still
think, that if parents could learn and would follow the
good advice which he gives, a great deal of discomfort and
trouble, and even of mortification, would be prevented.
Many persons have reason to regret the unwise forbearance
of parents who yielded to their reluctance, when children,
to visit the torturing room of the dentist. Dr. Woolworth's
precepts would diminish the necessity of painful operations,
and at the same time secure better results than are secured
by such operations, even when they are most skillfully performed.

THOMAS A. THACHER.
New Haven, May 17, 1869.

www.ingramcontent.com/pod-product-compliance
Lightning Source LLC
Chambersburg PA
CBHW021810190326
41518CB00007B/526